Snapwords
high-frequency picture words

MINI-LESSONS II

How to teach SnapWords®, integrating spelling, writing, and phonics concepts, with a focus on the structure of words & their spelling patterns

For SnapWords© Lists F, G, N1, N2, V

by Sarah Major, M.Ed.

Child1st's SnapWords® are 643 high-frequency words including the Dolch list of words, 300 Fry words, 500 Fountas & Pinnell words, and more. Each word has been carefully stylized to look like what it means in order to make learning high-frequency words a snap for the child.

On the reverse of each card is a suggested body motion to help active learners remember their sight words, and a sentence that uses the word correctly, leading to correct usage and reading comprehension.

The lessons in this book encompass the second set of 301 words, which include the Dolch words, and more...

Child1st Publications LLC

SnapWords® Mini-Lessons II

Ages 3 and up.

© 2019 Sarah Major
Printed in the United States of America

ISBN: 978-1-947484-09-2

Design: Sarah K Major

Published by Child1st Publications, LLC
800-881-0912 (phone)
888-886-1636 (fax)
www.child1st.com

All rights reserved. The reproduction of this book for any reason is strictly prohibited. No form of this work may be reproduced, transmitted or recorded without written permission from the publisher.

Other titles by Sarah Major:

The Right-Brained Math series: *Addition & Subtraction, Place Value, Multiplication & Division, and Fractions. The Illustrated Book of Sounds & Their Spelling Patterns, Alphabet Tales,* The Easy-for-Me™ Reading Program, the Easy-for-Me™ Children's Readers, and more.

Other educational work: Child1st SnapWords® and Alphabet.

TABLE OF CONTENTS

About this book 4

PART 1 Individual Mini-Lessons

List F: Level 1 5
         Level 2 12
         Level 3 16
         Level 4 20
         Level 5 24

List G: Level 1 28
         Level 2 33
         Level 3 38
         Level 4 43
         Level 5 47

List N1:   Level 1 51
         Level 2 56
         Level 3 60
         Level 4 64
         Level 5 68

List N2:   Level 1 72
         Level 2 77
         Level 3 81
         Level 4 85
         Level 5 89

List V: Level 1 93
         Level 2 98
         Level 3 103
         Level 4 108
         Level 5 112

PART 2 - Activities & Games 117
     ESOL Applications 118
     Word Recognition Activities 119
     Reading Activities 124
     Reading/Writing Connection Activities 126
     Structural Analysis of Words Activities 128
     Phonemic Awareness Activities 130

List of available SnapWords® 132

ABOUT THIS BOOK

SnapWords® Mini-Lessons II is one of those teaching tools you will reach for on a daily basis. There are mini-lessons for the second 301 SnapWords® that were designed. This list of 301 words includes words commonly used across the curriculum, nouns, and verbs.

Part 1 Mini-Lessons.
Each mini-lesson will help you guide the students into visual imprinting (committing the whole word with image into long term memory), writing the word, using it correctly in a sentence, and making sure the students comprehend the word in a sentence. Words that contain the same spelling patterns as the target word are included to expand learning.

Use with SnapWords® Lists F-V. The words in each list are provided for you in the front of each section of the book. The mini-lessons will guide you in the initial teaching/introduction of the words, and then you may use activities from the second part of the book to follow-up.

Part 2 *How to Teach SnapWords® High-Frequency Words using images, body motions, games and activities.*
This portion of the book includes activities for:
- word recognition
- word review
- words assessment
- oral sentence building
- tactile sentence building
- reading/writing connection
- story starters
- phonemic awareness
- sound spelling study
- word structure study

See more details on page 118.

Enjoy!

SNAPWORDS LIST F

List F Words - ABC order:	List F Words by level:
along also anything bed box car cat coat color dear dog door dress each early end face fat fine hand happy hat home hope later letter longer love maybe men money morning name night o'clock order pair part present push room sat second seem set sister someone something special stand store such thing third though until way yesterday yours	**Level 1:** cat, hat, fat, sat, bed, end, men, set, box, dog, car, way **Level 2:** dress, hand, stand, such, face, name, seem, coat, each, home, hope, fine **Level 3:** along, also, door, room, early, pair, part, store, dear, love, until, push **Level 4:** color, happy, later, letter, maybe, money, night, order, sister, thing, third, yours **Level 5:** anything, longer, morning, o'clock, present, second, someone, something, special, though, yesterday

How To Teach List F, Level 1

Show the children the SnapWord® card for CAT. Study the picture together and ask the children to comment on what they see. (Ex: The **c** is the face of the cat, the **a** is an orange tummy and there is a ragged tail rising out of the base of the **t**. Do the body motion from the reverse of the card. Now read the sentence from the back of the card to the children. Ask them to close their eyes and "see" the word and picture in their heads. Then ask them to open their eyes and write CAT on their whiteboards.

Write CAT on your whiteboard and brainstorm other words in the same family, writing these words in a column below CAT: hat, fat, sat, rat, mat, chat, that, etc.

Ask the children to make up sentences using cat and its related words. Have them choose their favorite to write and illustrate. The process of writing and drawing in this part of the lesson is very good for stimulating both hemispheres of the brain.

Show the children the SnapWord® card for HAT. Study the picture together and ask the children to comment on what they see. The picture shows a cowboy hat with the letters sitting on the hat band. The tall letters are on the outside, while the **a** is in between. Do the body motion and say the sentence from the back of the card. Ask the children to close their eyes until they can "see" the word and picture in their imagination. Next, have them write the word on their whiteboards. This part of the lesson is a routine that will be followed with every Mini-Lesson in this book: 1. Do the body motion 2. Say the sentence 3. Visualize the word and 4. Write the word on whiteboard.

Review the AT word family as in the lesson on CAT. Ask the children to make up sentences orally that use the word HAT. They will choose their favorite sentence to write and illustrate. EX: "I like my fuzzy hat."

List F, Level 1

fat

Show the children the SnapWord® card for FAT. Study the picture together and ask the children to comment on what they see. There is a yellow bird perched on the top of the **f**, the **a** is a very fat pig, and the **t** has a tree beside it. Follow the routine next: Body motion, sentence, visualization, and writing.

Review the AT family words. Ask the children to make up sentences using FAT and other AT words as desired. Write one and illustrate it.

sat

Show the children the SnapWord® card for SAT. Study the picture together and ask the children to comment on what they see. The **s** is a girl sitting on a green pillow looking happy to be there! On the other side of the **a** is a boy who is the **t**. He is waving a little flag and looking a bit upset that the girl is on his pillow. Follow the 4-step routine: Body motion, Sentence, Visualization, Writing.

Once again, review the AT family. Brainstorm some sentences that use as many of the AT family words as possible. Choose a sentence to write and illustrate.

bed

Show the children the SnapWord® card for BED. BED and the next three words all share a short E sound in the middle of the word. Study the picture together and ask the children to comment on what they see. The "bellies" of the **b** and **d** face each other with a **e** in the middle. The pillow is on the belly of the **b** while the slippers are by the **d**. The backs of the tall letters conveniently form the head and footboards of the bed. Note the picture right above the **e**. Follow the routine next: Body motion, sentence, visualization, and writing.

Write BED on your whiteboard and brainstorm other words that fit into this family: Ned, led, med, fed, Ted, wed.

Ask the children to make up sentences using bed, choose their favorite to write and illustrate.

List F, Level 1

end

Show the children the SnapWord® card for END. Study the picture together and ask the children to comment on what they see. The picture of END shows action! Each letter has a wheel under it, they are sitting on a platform and have just crashed into a brick wall! Notice that the back of the **d** is flat against the wall. The three letters are slightly different colors – the lightest is the **e** and the darkest is the **d**. Follow the routine next: Body motion, sentence, visualization, and writing.

Write END on your whiteboard and brainstorm other words that belong with it: bend, fend, send, tend, mend, vend, lend, pend, rend, blend, spend, trend, etc.

Ask the children to make up sentences using END, and then choose their best one to write and illustrate.

men

Show the children the SnapWord® card for MEN. Study the picture together and ask the children to comment on what they see. This picture is interesting. The **m** and the **n** are both men. Both men are checking the time and are not paying attention to each other. The **e** in the middle is just sitting there. Notice that the first man has a pocket watch and a long striped tie while the other man has a wrist watch and a little red tie. Follow the routine next: Body motion, sentence, visualization, and writing.

Write MEN on your whiteboard and add words from that family in a column. Ben, den, fen, hen, Jen, Ken, pen, ten, yen, then.

Children will make up a sentence they like, and then write and illustrate it.

set

Show the children the SnapWord® card for SET. Study the picture together and ask the children to comment on what they see. This picture is of a dining room. The **e** in the middle is the table which is covered with a lovely purple tablecloth. The **t** is a lady who is setting the table. She has one plate up in the air and several more in her other arm. Do you think the plates will all fit on the table? What color are her shoes? Follow the routine next: Body motion, sentence, visualization, and writing.

Write SET on your whiteboard and add words from that family in a column. Bet, get, jet, let, met, net, pet, vet, wet, yet.

Children will make up a sentence they like, and then write and illustrate it.

List F, Level 1

box

Show the children the SnapWord® card for BOX. Study the picture together and ask the children to comment on what they see. This picture is of a dining room. The **o** in the middle has a puppy peeking out while the **b** at the beginning has a nice straight back that follows the side of the box. The **x** fits nicely into the other corner. Follow the routine next: Body motion, sentence, visualization, and writing.

Write BOX on your whiteboard and add words from that family in a column. Fox, lox, pox. That's all, folks!

Children will make up a sentence they like, and then write and illustrate it.

dog

Show the children the SnapWord® card for DOG. Study the picture together and ask the children to comment on what they see. This picture is a bit floppy, isn't it? The **d** is the dog's head and there are two very large floppy ears on it and two bug eyes! The **o** is the tummy of the dog while the **g** is his back end. Note the floppy tail wagging! Follow the routine next: Body motion, sentence, visualization, and writing.

Write DOG on your whiteboard and add words from that family in a column. Bog, cog, fog, hog, jog, log, blog, clog, flog.

Children will make up a sentence they like, and then write and illustrate it.

car

Show the children the SnapWord® card for CAR. Study the picture together and ask the children to comment on what they see. The **c** in CAR is the back end of the car where the trunk is. The **a** conveniently is the middle part and has two windows above it. The **r** is the front of the car and the headlights are shining right out of the little hook in the **r**. Follow the routine next: Body motion, sentence, visualization, and writing. Focus on AR/R.

Write CAR on your whiteboard and add words from that family in a column. bar, far, jar, mar, par, tar, char, scar, spar, star.

Children will make up a sentence they like, and then write and illustrate it.

List F, Level 1

Show the children the SnapWord® card for WAY. Study the picture together and ask the children to comment on what they see. This picture is pretty cool. The **w** is actually a bendy road with a dotted line running down the middle. A really complicated sign with a vine growing up it and a bird sitting on top is just behind it. The **a** in the middle is just filled with green and is leaning on the **w**. The **y** is a girl who is very confused about which way to go! Follow the routine next: Body motion, sentence, visualization, and writing. Focus on AY/Ā.

Write WAY on your whiteboard and add words from that family in a column. bay, day, gay, hay, jay, lay, may, nay, pay, quay, ray, say, bray, clay, cray, ray, gray, play, pray, slay, spray, stay, stray, sway, tray. Whew!

Children will make up a sentence they like, and then write and illustrate it.

List F, Level 2

dress

Show the children the SnapWord® card for DRESS. Study the picture together and ask the children to comment on what they see. The **d** is the tall letter and is of course the lady wearing the dress. The **r** is back to back with the **d**. Note the double **s** at the end of the word.

Follow the routine next: Body motion, sentence, visualization, and writing.

Write DRESS on your whiteboard and add words from that family in a column. Less, mess, bless, chess, press, stress, tress.

Children will make up a sentence they like, and then write and illustrate it. Ex: "The dress was a mess, so Bess will press it." Or "I look like a mess so I will dress to play chess."

hand

Show the children the SnapWord® card for HAND. Study the picture together and ask the children to comment on what they see. This picture is interesting because the two tall letters, **h** and **d** are on the first and last fingers on the hand. Their "bellies" are facing each other so they can talk over the heads of the short letters (which also form the word AN). Have the child(ren) make their left hand into a fist, then as they stick out their pinky finger, they would sound "Hhhh", the middle fingers still curled down represent "A-N" and then as they stick out their thumb, they would say the sound of "dd." Point out also the smaller word "AND" inside of HAND.

Follow the routine next: Body motion, sentence, visualization, and writing.

Write HAND on your whiteboard and add words from that family in a column. And, band, land, sand, bland, brand, gland, stand, grand.

Children will make up a sentence they like, and then write and illustrate it. Ex: "Stand and give a hand to the best band in the land!" "I like both land and sand."

stand

Show the children the SnapWord® card for STAND. Study the picture together and ask the children to comment on what they see. Immediately we note that there are two tall letters, **t** and **d**, which are the dads of the three children. Do you think the **s**, the girl sitting on the green pillow, is the child of the **t** dad while the **an** children belong to the **d** dad? How do you think the children are feeling based on the looks on their faces? What do you think the dads are talking about?

Follow the routine next: Body motion, sentence, visualization, and writing.

Write STAND on your whiteboard point out that STAND belongs to the same word family as HAND, which they already learned.

Children will make up a sentence they like, and then write and illustrate it. Ex: "I like to stand in the sand." "I can stand on one hand." "Stand and give the band a hand."

List F, Level 2

such

Show the children the SnapWord® card for SUCH. Study the picture together and ask the children to comment on what they see. This picture tells a little story. The **s** girl is telling a whopping story and is all excited, waving her hands around. The word ends with CH and the girl on the **c** doesn't believe what she is hearing. She is saying "Never heard of such a thing!" What do you think the first girl is saying?

Follow the routine next: Body motion, sentence, visualization, and writing.

Write SUCH on your whiteboard and mention that SUCH is part of a very tiny family. It goes with the word MUCH. There are other words that rhyme with SUCH, but they are spelled differently: "Dutch, crutch, hutch, clutch." In these words, the digraph CH is spelled with a T: TCH.

Children will make up a sentence they like, and then write and illustrate it. Ex: "Much of what she said is such a good story!"

face

Show the children the SnapWord® card for FACE. Study the picture together and ask the children to comment on what they see. I love this picture. The **f** is perfect for the guy bending over to wash his face because the **f** is naturally curved over. The **a** is the sink, while the **c** is the towel rack. The **e** in this word goes with the **c** to make the **s** sound. It is busily pinching the **a** so she will say her name!

Follow the routine next: Body motion, sentence, visualization, and writing.

Write FACE on your whiteboard and add its family: "ace, lace, mace, pace, race, brace, grace, place, space, trace."

Children will make up a sentence they like, and then write and illustrate it. Ex: "She will race with lace about her face."

name

Show the children the SnapWord® card for NAME. Study the picture together and ask the children to comment on what they see. In this picture, we see two men shaking hands and introducing themselves. They are on the letters **n** and **m**. Each is wearing a red tie and we can see that the man on **m** has just written out a name tag that says, "Hello, my name is Joe."

Follow the routine next: Body motion, sentence, visualization, and writing.

Write NAME on your whiteboard and add its family: "came, dame, fame, game, lame, same, tame, blame, flame, frame, shame."

Children will make up a sentence they like, and then write and illustrate it. Ex: "The name of the game is lame." "The boys have the same name!"

List F, Level 2

seem

Show the children the SnapWord® card for SEEM. Study the picture together and ask the children to comment on what they see. This picture cracks me up. It tells a little story. The **s** lady went to get her hair colored and seems to be very upset at how it turned out. The stylist, the **m** lady, doesn't understand what is wrong! Doesn't she LIKE to have blue hair?? Note the scissors by the first **e** and the bottle of dye by the second one.

Follow the routine next: Body motion, sentence, visualization, and writing.

Write SEEM on your whiteboard and add its family: "deem, teem." (deem means to consider or judge – "I deem this to be an unsafe ride." Teem means to be full of something. "The pond teems with little fish." "The forest teems with life."

Children will make up a sentence they like, and then write and illustrate it. Ex: "I said to my dog, 'You seem to want more to eat." "The pond will seem to teem with fish."

each

Show the children the SnapWord® card for EACH. Study the picture together and ask the children to comment on what they see. This picture is cool. The three short letters, **e**, **a**, **c** is a small child. The tall letter **h** is a mom holding three suckers, one for each of the children. Note that all three short letters are rounded. Compare how they are formed. There are only two sounds in this word: Long **e** spelled **ea** and CH as in chip.

Follow the routine next: Body motion, sentence, visualization, and writing.

Write EACH on your whiteboard and add its family: "beach, leach, peach, reach, teach, bleach, breach, preach." Point out that each of these words have the word EACH in them. Just add a letter or two at the beginning of the word.

Children will make up a sentence they like, and then write and illustrate it. Ex: "Each of us will reach for a peach."

coat

Show the children the SnapWord® card for COAT. Study the picture together and ask the children to comment on what they see. The word COAT is written on sections of a coat. The **c** is the first sleeve, the **o** and **a** are two halves of the body of the coat, while the **t** is the right sleeve. Notice that the **o** and **a** are joined together by a big button. Focus on long **o** spelled **oa**.

Follow the routine next: Body motion, sentence, visualization, and writing.

Write COAT on your whiteboard and add its family: "boat, goat, moat, oat, bloat, gloat, float, throat."

Children will make up a sentence they like, and then write and illustrate it. Ex: "The goat will float on a boat in the moat."

List F, Level 2

home

Show the children the SnapWord® card for HOME. Study the picture together and ask the children to comment on what they see. The initial letter, **h** is the home, actually. The roof of the home is a mushroom cap, and the door to the home is inside the **h**. Notice the flower coming forward from inside the **m**. The home is just about the right size for Abner Beetle. In HOME and in HOPE (coming up next) the final **e** is a pinchy **e** who reaches around the consonant to pinch the vowel so that the vowel says its name.
   Follow the routine next: Body motion, sentence, visualization, and writing.
   Write HOME on your whiteboard and add its family: "dome, Rome, tome, gnome, chrome."
   Children will make up a sentence they like, and then write and illustrate it. Ex: "The home in Rome has a dome." "My home has a dome on top."

hope

Show the children the SnapWord® card for HOPE. Study the picture together and ask the children to comment on what they see. The **h** is a girl who is looking at the presents under the tree and she does hope one of them is for her! The presents are two rounded short letters: **o** and pinchy **e**. The Christmas tree is directly behind the **p** and there is a pretty blue ornament right in the center of the top of the **p**. Hope and home (last lesson) are almost exactly the same.
   Follow the routine next: Body motion, sentence, visualization, and writing.
   Write HOPE on your whiteboard and add its family: "cope, lope, mope, nope, pope, rope, grope, scope, slope."
   Children will make up a sentence they like, and then write and illustrate it. Ex: "I hope I can cope when you mope!" "I hope you will lope to get the rope!"

fine

Show the children the SnapWord® card for FINE. Study the picture together and ask the children to comment on what they see. Ed is a tall man shaped like an **f** while the lady is short and in the shape of an **i**. It looks like they met in a park and are chatting. The lady's flowers are spilling over onto the **n** and the final **e** is just sitting there innocently (but actually pinching the **i** to make her say her name!)
   Follow the routine next: Body motion, sentence, visualization, and writing.
   Write HOPE on your whiteboard and add its family: "dine, line, mine, nine, pine, tine, vine, wine, brine, shine, shrine, spine, swine, whine."
   Children will make up a sentence they like, and then write and illustrate it. Ex: "It is fine that the vine is on a line by the pine!" "Nine swine are in a line by the pine."

List F, Level 3

along

Show the children the SnapWord® card for ALONG. Study the picture together and ask the children to comment on what they see. Every letter in ALONG is important to the story. The mother is the tall **l** and she is looking back to ask the little pink girl (the letter **a**) if she wants to come along. The **o** is her large purse while the **n** and the **g** are the two children who are already coming along. Follow the routine next: Body motion, sentence, visualization, and writing.

Write ALONG on your whiteboard and add its family: "bong, dong, gong, long, song, tong, prong, strong, wrong."

Children will make up a sentence they like, and then write and illustrate it. Ex: "Along with the song there is a gong." "We have a long tong along with a strong prong."

also

Show the children the SnapWord® card for ALSO. Study the picture together and ask the children to comment on what they see. ALSO has two syllables. The picture shows them: the **a** and **l** are the first syllable, and the **so** of course is the second syllable. Note the tall fellow, **l**, has opened one present and he want the large orange one also!

Follow the routine next: Body motion, sentence, visualization, and writing.

Write ALSO on your whiteboard. ALSO has a very tiny word family: "torso, corso." (Corso is an Italian word meaning road or street.)

Children will make up a sentence they like, and then write and illustrate it. Ex: "His legs are long and also his torso!"

door

Show the children the SnapWord® card for DOOR. Study the picture together and ask the children to comment on what they see. I like the picture for DOOR because the **oo** is two letters working together to make the OR sound with the **r**. Notice that they also are two parts of one door and each one has a gold door knob.

Follow the routine next: Body motion, sentence, visualization, and writing.

Write DOOR on your whiteboard and add its family: "poor, boor, moor spoor."

Children will make up a sentence they like, and then write and illustrate it. Ex: "The poor boor sits by the door to the moor."

List F, Level 3

room

Show the children the SnapWord® card for ROOM. Study the picture together and ask the children to comment on what they see. Notice that the **r** is holding a reading lamp over the bed. Next are two windows made up of the two **o**'s that work together to make the long OO sound. Finally, the **m** is two blue closet doors sitting between the windows and the dresser.

Follow the routine next: Body motion, sentence, visualization, and writing.

Write ROOM on your whiteboard and add its family: "boom, doom, loom, zoom, bloom, broom, gloom, groom."

Children will make up a sentence they like, and then write and illustrate it. Ex: "I zoom to my room when I hear a boom!" "There is room for the bloom by the broom."

dear

Show the children the SnapWord® card for DEAR. Study the picture together and ask the children to comment on what they see. This picture has a mother who is the tall **d**, and she has three children who are dear to her. The **e** child has reddish hair, the **a** child has short hair and a green shirt, and the **r** child has a ponytail and is dragging a blankie behind her.

Follow the routine next: Body motion, sentence, visualization, and writing.

Write DEAR on your whiteboard and add its family: "ear, fear, gear, hear, near, rear, sear, tear, year, clear, shear, smear, spear."

Children will make up a sentence they like, and then write and illustrate it. Ex: "I fear my dear cats are near the rear." "Dear, I hear your gear is in the rear of the car."

pair

Show the children the SnapWord® card for PAIR. Study the picture together and ask the children to comment on what they see. This picture shows several things that make a pair: A pair of socks, dice, pants, and mittens. The **i** is a girl talking about the pairs she has. Notice that the oval green rug is under the short letters and under the rounded part of the **p**. The girl's mittens are over the **a** and the **r**. Focus on small word "air".

Follow the routine next: Body motion, sentence, visualization, and writing.

Write PAIR on your whiteboard and add its family: "air, fair, hair, lair, chair, flair, stair."

Children will make up a sentence they like, and then write and illustrate it. Ex: "My pair is in the air!" "At the fair, the chair sat in the lair by the chair."

List F, Level 3

part

Show the children the SnapWord® card for PART. Study the picture together and ask the children to comment on what they see. PART means two things. A slice of pizza is part of the whole pizza. There is also a part in someone's hair. In this picture, two boys are combing their hair and one boy is showing off his perfect part. The **p** boy with the green comb says, "Wowzers!" when he sees it. The **a** and **r** work together to say AR and they are leaning back also admiring the part. **t** is the boy who made the amazing part which he made with his red comb.

Follow the routine next: Body motion, sentence, visualization, and writing.

Write PART on your whiteboard and add its family: "Bart, cart, dart, mart, tart, chart, smart, start."

Children will make up a sentence they like, and then write and illustrate it. Ex: "I put part of it in the cart at the mart." "Part of the apple is tart."

store

Show the children the SnapWord® card for STORE. Study the picture together and ask the children to comment on what they see. In this picture, the store has a pretty green awning. Right under it is a red sign that says "candy." Right under the red sign is **s**. The door has the **t** on it. The **o** is the little girl who is looking in the window at the candy. Finally we have **re**. The e at the end is just hanging out with the **or**, helping them say OR. Notice that all the short letters have a yellow wall behind them.

Follow the routine next: Body motion, sentence, visualization, and writing.

Write STORE on your whiteboard and add its family: "bore, core, fore, gore, more, pore, sore, tore, wore, chore, score, shore, snore, spore, swore."

Children will make up a sentence they like, and then write and illustrate it. Ex: "I fell at the store and tore what I wore!" "The store by the shore will score more!"

early

Show the children the SnapWord® card for EARLY. Study the picture together, noticing every little detail, and ask the children to comment on what they see. This picture makes me sleepy. I can almost hear Mr. **e** snoring, can't you? The **ear** letters are on the bed together because they work with each other to make the sound ER. The tall l holds a reading lamp over the end of the bed, while the **y** in front of the window holds the blue robe. If you look at the clock you can see it is 4:00 in the morning! That is very early, isn't it?

Follow the routine next: Body motion, sentence, visualization, and writing.

Write EARLY on your whiteboard and add: "pearly, earth learn, heard."

Children will make up a sentence they like, and then write and illustrate it. Ex: "The early sky is pearly." "I got up early to learn about the earth."

List F, Level 3

love

Show the children the SnapWord® card for LOVE. Study the picture together and ask the children to comment on what they see. The **l** is a girl who apparently just got a rose for valentine's day. The **v** is the big heart that just overlaps the **o** and the **e**. In this word, the o says "UH", the sound that short **u** makes. The final **ve** work together to say the sound **v** normally makes on his own.

Follow the routine next: Body motion, sentence, visualization, and writing.

Write LOVE on your whiteboard and add its family: "dove, glove, shove." There are other words that are spelled the same at the end of the words, but they don't sound like LOVE. (cove, wove, clove, etc.) In these words, the final e is a pinchy **e** making the **o** a long **o** sound.

Children will make up a sentence they like, and then write and illustrate it. Ex: "I love my dove." "My dove will shove the glove that I love."

until

Show the children the SnapWord® card for UNTIL. Study the picture together and ask the children to comment on what they see. This word has two syllables: UN and TIL. The UN portion of this picture shows a huge mess in the boy's room. Notice that the **u** is a trashcan that is overflowing. The **n** is a nightstand that is covered with stuff. The **t** is a boy who is dismayed because his mother is saying he can't play until he cleans up. We can see the **l** is the mother who is talking to him, and she is standing beside the **l** which could have been a broom or a mop.

Follow the routine next: Body motion, sentence, visualization, and writing.

Write UNTIL on your whiteboard and add related words: "anvil, basil, Brazil, civil, fossil, gerbil, pupil, pistil, pencil, nostril."

Children will make up a sentence they like, and then write and illustrate it. Ex: "I will try until I have a civil gerbil." "Does basil in Brazil grow until it has a pistil?"

push

Show the children the SnapWord® card for PUSH. Study the picture together and ask the children to comment on what they see. **P** and **u** are guys helping the **SH** push a car which has stalled. It is pretty funny that they have formed a line and are pushing each other and only one boy is pushing the car.

Follow the routine next: Body motion, sentence, visualization, and writing.

Write PUSH on your whiteboard and add its family: "bush, ambush, rosebush, firebush." Many words end with USH but they don't sound like PUSH. (Ex: gush, hush, lush, mush, etc.)

Children will make up a sentence they like, and then write and illustrate it. Ex: "We will push until the bush falls over." "I will push for a rosebush and firebush in the yard."

List F, Level 4

color

Show the children the SnapWord® card for COLOR. Study the picture together and ask the children to comment on what they see. This picture shows that each letter in the word COLOR has its own color. Note that the tall **l** is in the middle on blue, the two **o**s are on either side on orange and yellow. The **c** leads with green, while the **r** ends with red.

Follow the routine next: Body motion, sentence, visualization, and writing.

Write COLOR on your whiteboard and add its family: "actor, author, sailor, editor, vendor, minor, doctor."

Children will make up a sentence they like, and then write and illustrate it. Ex: "He will color the actor, the doctor, and the sailor blue." The minor author will color it all red."

happy

Show the children the SnapWord® card for HAPPY. Study the picture together and ask the children to comment on what they see. This picture is happy! It is someone's birthday and three girls are happy to be singing to their friend. Notice the first girl in purple who is an **h**. The **a** is a pink-dotted present. The first **p** is the girl in the middle who is wearing yellow and waving both hands. The next **p** is not decorated, but sounds along with the first **p**. The third girl is a **y** and she is wearing a blue scarf and shoes.

Follow the routine next: Body motion, sentence, visualization, and writing.

Write HAPPY on your whiteboard and add its family: "flappy, sappy, pappy, tappy, scrappy."

Children will make up a sentence they like, and then write and illustrate it. Ex: "My pappy is scrappy but also happy." "It is not sappy to be happy!"

letter

Show the children the SnapWord® card for LETTER. Study the picture together and ask the children to comment on what they see. This picture shows two different meanings for the word LETTER. There is a letter in the **l** mailbox. The letter has some letters on it. Notice that the mailbox flap is resting on the **e**. Next are two **t**'s and the girl is the second one. The **er** at the end work together to make their sound. Notice the pink flowers curling up over the **e**. The structure of this word is interesting because if you divide it in half between the Ts, the next letter out is an E in both halves. Then you have the L and R at each end of the word.

Follow the routine next: Body motion, sentence, visualization, and writing.

Write LETTER on your whiteboard and add its family: "better, setter, getter, fetter, wetter."

Children will make up a sentence they like, and then write and illustrate it. Ex: "The letter got wetter." "The setter got my letter wetter."

List F, Level 4

later

Show the children the SnapWord® card for LATER. Study the picture together and ask the children to comment on what they see. Two ladies are talking in this picture. They are the tall letters: **l** and **t**. Between them is the short **a**. At the end of the word, **er** work together to make their ending sound.

Follow the routine next: Body motion, sentence, visualization, and writing.

Write LATER on your whiteboard and add its family: "cater, gater, hater, tater, rater, waiter."

Children will make up a sentence they like, and then write and illustrate it. Ex: "Later, the waiter will cater the tater tots." "The rater is a hater of the later tater tots."

maybe

Show the children the SnapWord® card for MAYBE. Study the picture together and ask the children to comment on what they see. This picture is in the Spring when rains spring up suddenly from gray clouds. The word is a compound word. MAY has an **a** which is a pond full of blue water. The **b** is a girl holding an umbrella and thinking that maybe later it will stop raining so she can go swimming in the blue pond. The final **e** is holding her swimsuit and towel.

Follow the routine next: Body motion, sentence, visualization, and writing.

Write MAYBE on your whiteboard and add: "he, she, we."

Children will make up a sentence they like, and then write and illustrate it. Ex: "Maybe he and she will run."

money

Show the children the SnapWord® card for MONEY. Study the picture together and ask the children to comment on what they see. This picture makes an impact because money is falling all over the place! The **o** is a bag full of money – but not falling money like the **y** is. That bag ripped open and lots of money fell out onto the floor. The **o** in money sounds like a short **u** so seeing the round bag of money will help children remember to spell money with an **o**. The **e** almost looks like he is surprised at the money rolling out of the **y**.

Follow the routine next: Body motion, sentence, visualization, and writing.

Write MONEY on your whiteboard and add its family: "honey, boney, alley, nosey, mopey, trolley, holey."

Children will make up a sentence they like, and then write and illustrate it. Ex: "Honey, my bag was holey and the money fell into the alley." "The money bag was holey."

List F, Level 4

night

Show the children the SnapWord® card for NIGHT. Study the picture together and ask the children to comment on what they see. The picture for NIGHT is very helpful because the first two sounds **ni**, are bathed in moonlight, then right after the silent **gh**, a little tree separates those letters from the final **t**.

Follow the routine next: Body motion, sentence, visualization, and writing.

Write NIGHT on your whiteboard and add its family: knight, light, might, right, sight, tight, blight, bright, flight, fright, plight, slight.

Children will make up a sentence they like, and then write and illustrate it. Ex: "The moon is a bright light in the night." or "The bright light of the moon is a sight in the night!"

order

Show the children the SnapWord® card for ORDER. Study the picture together and ask the children to comment on what they see. This picture makes me think of summer at the beach. The **or** are a couple of girls placing an order for hotdogs. Inside the food stand is a tall **d** man who is taking their order. The platter of hotdogs is resting over the **e** in **er**. Notice that the green wall also separates the two syllables in the word.

Follow the routine next: Body motion, sentence, visualization, and writing.

Write ORDER on your whiteboard and add: "border, finder, glider, fender, older, elder, gander, alder, rider, under, hider, wader, wider."

Children will make up a sentence they like, and then write and illustrate it. Ex: "The rider will order an older glider with a fender." "The order had a gander under the border."

sister

Show the children the SnapWord® card for SISTER. Study the picture together and ask the children to comment on what they see. This word has two syllables and in this picture, the syllables are divided between the two girls. The second **s** is the baby sister while the **t** is the older sister who is reading. **er** is a common ending and we have seen it before in LETTER, LATER, ORDER, and now SISTER.

Follow the routine next: Body motion, sentence, visualization, and writing.

Write SISTER on your whiteboard and refer to the list of words for ORDER.

Children will make up a sentence they like, and then write and illustrate it. Ex: "My sister will order a gander for dinner."

List F, Level 4

thing

Show the children the SnapWord® card for THING. Study the picture together and ask the children to comment on what they see. The **th** that starts this word is a purple contraption (thing) that the boy, who is the **i** is making. The **n** is all tangled up in the cord he is going to plug in. His friend, the **g**, is shocked and surprised at the THING he has made.

Follow the routine next: Body motion, sentence, visualization, and writing.

Write THING on your whiteboard and add its family: "ding, bring, fling, king, sling, string, ring, sting, etc."

Children will make up a sentence they like, and then write and illustrate it. Ex: "The thing I saw has a string with a ring to pull." "The king will fling the thing."

third

Show the children the SnapWord® card for THIRD. Study the picture together and ask the children to comment on what they see. The girl in this word is the **i**. No matter which way you look, she is the third one! Three of the letters are tall: **th** and the final **d**. The two short letters in the middle say ER together (spelled IR).

Follow the routine next: Body motion, sentence, visualization, and writing.

Write THIRD on your whiteboard and add: "bird, gird, dirt, girl, sir skirt shirt fir." (gird means to circle oneself like you do when you put on a belt. It also means to prepare oneself for something difficult.)

Children will make up a sentence they like, and then write and illustrate it. Ex: "The third bird is in the dirt."

yours

Show the children the SnapWord® card for YOURS. Study the picture together and ask the children to comment on what they see. This word is very interesting. There are four words that speak of belonging: "our", "ours", "your", and "yours". The **y** at the beginning is the girl who is giving a plant to the boy, saying "This is yours." The **o** contains water to use when potting plants. The **u** contains the needed soil. The **r** is the boy who is receiving the plant and the **s** is the perch for a bluebird.

Follow the routine next: Body motion, sentence, visualization, and writing.

Write YOURS on your whiteboard. The words we add will have the sound spelling OUR when it sounds like OR. "fours, pours."

Children will make up a sentence they like, and then write and illustrate it. Ex: "I see that yours pours."

List F, Level 5

anything

Show the children the SnapWord® card for ANYTHING. Study the picture together and ask the children to comment on what they see. The focus of this picture is interesting. Anything is a compound word and your child(ren) have studied both of the individual words already: ANY and THING.

Follow the routine next: Body motion, sentence, visualization, and writing.

Write ANYTHING on your whiteboard. Then add other words that end with "thing." Thing, nothing, something, everything, clothing, bathing, soothing, seething, plaything."

Children will make up a sentence they like, and then write and illustrate it. Ex: "Anything is better than nothing." "Anything can be a plaything."

longer

Show the children the SnapWord® card for LONGER. Study the picture together and ask the children to comment on what they see. This word is all about comparisons. He has something. She has something long. You have something longer. You have already taught LONG, so notice that the rulers are by that word and the ER is extra. The girl who is the **r** is noticing that the **l** is longer than its ruler!

Follow the routine next: Body motion, sentence, visualization, and writing.

Write LONGER on your whiteboard. Then add: "bong, dong, gong, song, tong, strong, wrong."

Children will make up a sentence they like, and then write and illustrate it. Ex: "This tong is stronger and longer." "The song was longer than I wanted."

morning

Show the children the SnapWord® card for MORNING. Study the picture together and ask the children to comment on what they see. Any word that ends in ING can be broken down just by taking off the ending. What is left here is MORN. Inside MORN is a little word OR. Notice the white bunny under the **o**. The rising sun is behind the **n** and the happy girl is the **i**. She also starts the second syllable in the word.

Follow the routine next: Body motion, sentence, visualization, and writing.

Write MORNING on your whiteboard. Add: "horn, corn, adorn, lorn, torn, worn" – all words with OR inside them.

Children will make up a sentence they like, and then write and illustrate it. Ex: "In the morning I hear a worn horn."

List F, Level 5

o'clock

Show the children the SnapWord® card for O'CLOCK. Study the picture together and ask the children to comment on what they see. Immediately we notice that there are two clocks – each is an **o**. There is a spider and web right near where the apostrophe falls in the word. Notice the mouse sitting on the second **o**. This word might be funny, but as long as you remember to include the apostrophe, you can basically sound it out!

Follow the routine next: Body motion, sentence, visualization, and writing.

Write "O'CLOCK on your whiteboard. Then add: "block, smock, stock, mock, tock, chock, jock, shock."

Children will make up a sentence they like, and then write and illustrate it. Ex: "At 2 o'clock I hear a tick and a tock." "At 2 o'clock we took stock."

present

Show the children the SnapWord® card for PRESENT. Study the picture together and ask the children to comment on what they see. Present means two different things: it can be a gift and it also means you are here when someone is taking roll. Notice that the first **p** is a girl yelling "I'm present!" when the teacher calls her name. Notice also the two presents that are both shaped like an **e**. Three other children are on the **r**, **s**, and the **n**. The two syllables are PRE-SENT.

Follow the routine next: Body motion, sentence, visualization, and writing.

Write PRESENT on your whiteboard. Add other words that rhyme: "bent, spent, advent, lent, went, dent, rent, vent, tent."

Children will make up a sentence they like, and then write and illustrate it. Ex: "The present is a tent." "I spent my money on a tent with a vent."

second

Show the children the SnapWord® card for SECOND. Study the picture together and ask the children to comment on what they see. This word picture is cool because you can see it rounding the bases! It puts you right into a baseball game! There are two syllables: SE-COND and the **o** is the very spot where second base falls. The only tall letter in the word is the **d** – the guy running from first to second base.

Follow the routine next: Body motion, sentence, visualization, and writing.

Write SECOND on your whiteboard. While there are several words that end with OND, most of them sound like POND – not the way we say SECOND. Here they are: "pond, blond, fond, almond, beyond, respond."

Children will make up a sentence they like, and then write and illustrate it. Ex: "The second blond is in the pond."

List F, Level 5

someone

Show the children the SnapWord® card for SOMEONE. Study the picture together and ask the children to comment on what they see. There are two known words in SOMEONE. SOME and ONE. These girls are each in one of the words and boy are they yelling loudly! There is a huge green snake very close by and they are yelling for someone to help them! Nothing that **e** is of very little help as both of them are just throwing their hands up and hollering for help!

Follow the routine next: Body motion, sentence, visualization, and writing.

Write SOMEONE on your whiteboard. Let's add other words: "anyone, no one, done, ton, undone."

Children will make up a sentence they like, and then write and illustrate it. Ex: "Someone has done this." "Someone needs to help. Will anyone?"

something

Show the children the SnapWord® card for SOMETHING. Study the picture together and ask the children to comment on what they see. There are two known words: SOME and THING. The poor guy who is the **t** wants to make something out of all the stuff on his floor, but he's just not sure how to do it! Notice that he is the first letter in the second word.

Follow the routine next: Body motion, sentence, visualization, and writing.

Write SOMETHING on your whiteboard. Refer to the list of words in the lessons for thing and anything.

Children will make up a sentence they like, and then write and illustrate it. Ex: "I want to make something...anything!" "This thing will be something cool soon!"

special

Show the children the SnapWord® card for SPECIAL. Study the picture together and ask the children to comment on what they see. The word SPECIAL could be tricky to remember how to spell. This is where the picture will be helpful! You can hear the **s** at the beginning. Then the **p** and the **e** are two girls. SPE so far! Next we have the sound of SH, but it is spelled CI. Note that the boy is the large **i** in this word. Finally you can hear the AL ending. You can remember this spelling by remembering the guy who thought he was very special so he wore a golden crown.

Follow the routine next: Body motion, sentence, visualization, and writing.

Write SPECIAL on your whiteboard. Let's add other words: "social, crucial."

Children will make up a sentence they like, and then write and illustrate it. Ex: "The social was special." "It is crucial to have a special social time."

List F, Level 5

though

Show the children the SnapWord® card for THOUGH. Study the picture together and ask the children to comment on what they see. This word is very funny because it has a lot of letters, but only two sounds! Talk about the **ough** spelling which can be found in many words. There are two short, rounded letters followed by two letters, one with a tail and the other a tall letter.

Follow the routine next: Body motion, sentence, visualization, and writing.

Write SOMEONE on your whiteboard. Let's add other words: "dough, although, furlough."

Children will make up a sentence they like, and then write and illustrate it. Ex: "We ate it though the dough was cold."

yesterday

Show the children the SnapWord® card for YESTERDAY. Study the picture together and ask the children to comment on what they see. This word has three syllables and in this picture you can see them clearly. The picture also shows the meaning of YESTERDAY. YES and DAY are complete words, while TER rhymes with HER.

Follow the routine next: Body motion, sentence, visualization, and writing.

Write YESTERDAY on your whiteboard separating the word into syllables. Next children will make up a sentence they like, and then write and illustrate it. Ex: "Yesterday was the day before this day."

SNAPWORDS LIST G

List G Words - ABC order:	List G Words by level:
able above against almost already although among bad beautiful behind below between brother certain dark deep dry during easy either else ever favorite few finally free front fun great half heavy important inside instead large less lot mad main nice often page perhaps possible probably quick ready real really scared several sick side simple since size sound sure that's themselves they're top	**Level 1:** able, bad, dry, else, ever, fun, less, lot, mad, sick, top, that's **Level 2:** dark, deep, easy, few, free, half, main, nice, page, real, side, size, sure **Level 3:** above, among, below, front, great, heavy, large, often, quick, ready, really, since, sound **Level 4:** against, almost, already, behind, certain, during, either, finally, inside, scared, simple, they're **Level 5:** although, beautiful, between, brother, favorite, important, instead, perhaps, possible, probably, several, themselves,

How To Teach   List G, Level 1

Show the children the SnapWord® card for ABLE. Study the picture together and ask the children to comment on what they see. (Ex: there are two children with untied shoes. They are the short letters, **a** and **e**. One child feels ABLE to tie his shoes while the other doesn't feel ABLE)

Do the body motion from the reverse of the card. Now read the sentence from the back of the card to the children. Ask them to close their eyes and "see" the word and picture in their heads. Then ask them to open their eyes and write ABLE on their whiteboards. Note that LE at the end says /L/.

Write ABLE on your whiteboard and brainstorm other words in the same family, writing these words in a column below ABLE: fable, cable, table, Mable, sable, etc.

Ask the children to make up sentences using ABLE and its related words. Have them choose their favorite to write and illustrate. The process of writing and drawing in this part of the lesson is very good for stimulating both hemispheres of the brain. Idea: "Mable tied a cable to the table."

Show the children the SnapWord® card for BAD. Study the picture together and ask the children to comment on what they see. The picture shows a child in a bad mood. The tall letters (**b** and **d**) are wondering why he is in a bad mood.

Do the body motion and say the sentence from the back of the card. Ask the children to close their eyes until they can "see" the word and picture in their imagination. Next, have them write the word on their whiteboards. This part of the lesson is a routine that will be followed with every Mini-Lesson in this book: 1. Do the body motion 2. Say the sentence 3. Visualize the word and 4. Write the word on whiteboard.

Write BAD on your whiteboard. The word family for this lesson is AD. Other words that go along with BAD are: cad, fad, had, lad, sad, and tad.

They will make up a sentence to write and illustrate. EX: "The lad is a tad sad."

dry

Show the children the SnapWord® card for DRY. Study the picture together and ask the children to comment on what they see. Two kids have just gone swimming in a pond and are wet. They have one towel to dry off with.

Follow the routine next: Body motion, sentence, visualization, and writing.

Write DRY on your whiteboard. Words ending in Y when it sounds like long I are: by, fry, cry, try, pry, my.

Write a sentence and illustrate it. "I won't cry when I try and fry my onions!"

List G, Level 1

else

 Show the children the SnapWord® card for ELSE. Study the picture together and ask the children to comment on what they see. The tall letter, **l**, is supervising while **s** packs a bag which is the final **e**. He is asking "What else do you need?
 Follow the 4-step routine: body motion, sentence, visualization, writing.
 Write ELSE on your whiteboard. The sound spelling focus for ELSE is the final SE which sounds like an S. Other words that follow this pattern are: house, mouse, horse, goose, nurse, sense, tense, etc.
 Write and illustrate a sentence. Ex: "The horse, mouse, and goose ran around the house."

ever

 Show the children the SnapWord® card for EVER. Study the picture together and ask the children to comment on what they see. Two children are in line to buy ice cream. The girl in the front seems to be taking a long time to find her money to pay. The child behind her is dramatically crying, "Will she ever finish?"
 Follow the routine next: body motion, sentence, visualization, and writing.
 Write EVER on your whiteboard and brainstorm other words that fit into this family: never, lever, sever, over, river, sliver.
 Ask the children to make up sentences using ever and then choose their favorite to write and illustrate. Ex: "We never ever go over the river."

fun

Show the children the SnapWord® card for FUN. Study the picture together and ask the children to comment on what they see. This poor guy doesn't think he is having fun! He is washing dishes and maybe his friends are playing. Maybe he will learn that washing dishes doesn't take long and can be fun!

Follow the routine next: body motion, sentence, visualization, and writing.

Write FUN on your whiteboard and brainstorm other words that belong with it: bun, gun, nun, pun, run, sun, shun, spun, stun.

Ask the children to make up sentences using FUN and then choose their best one to write and illustrate. Ex: "It was fun to run in the sun."

List G, Level 1

less

Show the children the SnapWord® card for LESS. Study the picture together and ask the children to comment on what they see. The l is a girl who has a big bowl of pink ice cream. The s is a boy with a smaller bowl of pink ice cream. He has LESS than she does.

Follow the routine next: body motion, sentence, visualization, and writing.

Write LESS on your whiteboard and add words from that family in a column. mess, bless, chess, dress, press, stress, tress.

Children will make up a sentence they like, and then write and illustrate it. Ex: "She will press the dress so she can play chess."

lot

Show the children the SnapWord® card for LOT. Study the picture together and ask the children to comment on what they see. This picture shows two meanings of LOT. There are a LOT of rocks in the empty LOT.

Follow the routine next: body motion, sentence, visualization, and writing.

Write LOT on your whiteboard and add words from that family in a column. cot, dot, got, hot, jot, knot, not, pot, rot, tot, blot, clot, plot, shot, slot, spot, trot.

Children will make up a sentence they like, and then write and illustrate it. Ex: "Dot got hot in the cot."

mad

Show the children the SnapWord® card for MAD. Study the picture together and ask the children to comment on what they see. In this picture it is easy to see "mad!" Sarge is stomping his foot and waving his fists.

Follow the routine next: body motion, sentence, visualization, and writing.

Write MAD on your whiteboard and add words from that family in a column. fad, dad, had, lad, bad, pad, sad, tad, clad, glad.

Children will make up a sentence they like, and then write and illustrate it. Ex: "I'm glad the lad is not mad or sad!"

List G, Level 1

sick

Show the children the SnapWord® card for SICK. Study the picture together and ask the children to comment on what they see. In this picture, the poor **i** is a child who is sick. He/she is sneezing and has a tissue ready to use on his/her nose. The sneeze was so big, the **ck** are tipping a bit!

Follow the routine next: body motion, sentence, visualization, and writing.

Write SICK on your whiteboard and add words from that family in a column. kick, lick, pick, quick, tick, wisk, brick, chick, click, flick, slick, stick, thick, trick.

Children will make up a sentence they like, and then write and illustrate it. Ex: "The chick is quick to pick a thick stick."

top

Show the children the SnapWord® card for TOP. Study the picture together and ask the children to comment on what they see. YUM! We see an ice cream cone with a cherry on top. The cherry is right over the **o** in top.

Follow the routine next: body motion, sentence, visualization, and writing.

Write TOP on your whiteboard and add words from that family in a column. bop, cop, hop, mop, pop, sop, chop, crop, drop, flop, plop, prop, shop, slop, stop.

Children will make up a sentence they like, and then write and illustrate it. Ex: "I saw the crop drop in the shop so I will stop and get a mop."

that's

Show the children the SnapWord® card for THAT'S. Study the picture together and ask the children to comment on what they see. This picture is pretty cool. We can see the boys who are the **t** and the **a** looking startled because of the boot (the apostrophe) kicking the "i" out of the word IS. THAT'S is made of the two words, THAT & IS. The apostrophe is placed where the "i" is missing.

Follow the routine next: body motion, sentence, visualization, and writing.

Write THAT'S on your whiteboard. If desired, work through the process of starting with the two words THAT, IS and making a contraction with them. Other words to use include: WHAT, IS > WHAT'S. HE, IS > HE'S. SHE, IS > SHE'S.

Children will make up a sentence they like, and then write and illustrate it. Ex: "SHE'S able to see THAT'S not going to work."

List G, Level 2

Show the children the SnapWord® card for DARK. Study the picture together and ask the children to comment on what they see. The two tall letters are the parents while the a is a child who doesn't seem to be excited about going into the dark cave! Mom, actually is whipping her head around to see if there is a better place to go. Do you think Dad is thrilled about going into the cave? Emphasize the fact that **ar** in the middle of the word say the name of the letter **r**.

Follow the routine next: Body motion, sentence, visualization, and writing.

Write DARK on your whiteboard and add words from that family in a column. The sound spelling on which to focus will be the AR spelling. Art, cart, park, lark, mark, star, bar, tar, etc. Ask the children to highlight or underline the AR in each of the words. You may include other words that are longer as desired: chart, spark, stark, alarm, party.

Children will make up a sentence they like, and then write and illustrate it. Ex: "I saw a star in the dark at the park."

Show the children the SnapWord® card for DEEP. Study the picture together and ask the children to comment on what they see. Two children (the **d** and one of the **e**'s) are staring down into a deep well. One of the children is about to drop a rock into the well. The sound spelling focus for DEEP is the **ee** which is a long **e** spelling.

Follow the routine next: Body motion, sentence, visualization, and writing.

Write DEEP on your whiteboard and add words from that family in a column. Steep, peep, seep, sheep, see, seed, cheese, wheels, tree, green.

Children will make up a sentence they like, and then write and illustrate it. Ex: "I see a seed in the deep hole by the tree."

easy

Show the children the SnapWord® card for EASY. Study the picture together and ask the children to comment on what they see. This picture shows a teacher writing an EASY problem on the board: 1+1. We made the e into the student to help point out that you hear a long **e** sound at the beginning of the word. You also hear a long **e** sound at the end of the word, but this one is spelled with a **y** pretending to be an **e**. The sound spelling focus is on **ea** sounding like long **e**.

Follow the routine next: Body motion, sentence, visualization, and writing.

Write EASY on your whiteboard and add other words with the EA / long E sound spelling: eat, seat, beach, leap, heap, neat, meat.

Children will make up a sentence they like, and then write and illustrate it. Ex: "It is easy to eat on a seat at the beach."

List G, Level 2

few

Show the children the SnapWord® card for FEW. Study the picture together and ask the children to comment on what they see. In this picture, there is a girl holding a large jar with just a few marbles in it. All the rest of the marbles from the jar have covered the floor. She is saying she is going to just keep a few of the marbles.

Follow the routine next: Body motion, sentence, visualization, and writing.

Write FEW on your whiteboard and point out the sound spelling EW. Other words that belong with FEW include: dew, grew, flew, chew, brew, crew, stew, threw, blew.

Children will make up a sentence they like, and then write and illustrate it. Ex: "I saw dew on the stew so I threw it out!" Or "The baby birds grew and then flew away."

free

Show the children the SnapWord® card for FREE. Study the picture together and ask the children to comment on what they see. In this picture, the **ee** spelling is two pet dogs. Their master is going to free them so they can play in the doggie park.

Follow the routine next: Body motion, sentence, visualization, and writing.

Write FREE on your whiteboard and add its family: see, tree, bee, spree, and other words from the lesson for DEEP.

Children will make up a sentence they like, and then write and illustrate it. Ex: "I see a bee in the tree." Or "You are free to climb into the green tree."

half

Show the children the SnapWord® card for HALF. Study the picture together and ask the children to comment on what they see. I hope that the girl in this picture is careful as she cuts the cake in half. The knife she is holding looks big! The word HALF has three tall letters. We can hear the **h** and the final **f**, but not the **l**. It will help if you encourage the children to pronounce the word in such a way that they can hear every letter.

Follow the routine next: Body motion, sentence, visualization, and writing.

Write HALF on your whiteboard and add its tiny family: calf. Other words that have a "silent" L include: walk, talk, chalk, could, would, should.

Children will make up a sentence they like, and then write and illustrate it. Ex: "We each washed half of the calf." "I walk and talk with chalk."

List G, Level 2

main

    Show the children the SnapWord® card for MAIN. Study the picture together and ask the children to comment on what they see. In this picture, two boys are talking about Main Street. One boy is the **i** with the dot being his mouth. He's saying, "Main Street is the main street in town." Today focus on long **a** spelled **ai**.
    Follow the routine next: Body motion, sentence, visualization, and writing.
    Write MAIN on your whiteboard and add its family: pain, mail, snail, nail, pail, frail, tail, hail, rain, stain, train.
    Children will make up a sentence they like, and then write and illustrate it. Ex: "I put a snail in the main pail."

nice

    Show the children the SnapWord® card for NICE. Study the picture together and ask the children to comment on what they see. This picture shows "nice mice!" Of course the boy with his wide open mouth is the lowercase **i**. The ending of the word which sounds like an **s** is spelled **ce**. This is the focus spelling pattern.
    Follow the routine next: Body motion, sentence, visualization, and writing.
    Write NICE on your whiteboard and add its family: ice, mice, lice, rice, spice.
    Children will make up a sentence they like, and then write and illustrate it. Ex: "Mice like some nice rice with spice."

page

    Show the children the SnapWord® card for PAGE. Study the picture together and ask the children to comment on what they see. In this picture, four children are sharing a book with two children on each page. Each letter is a child. The focus sound spelling is the final **ge** which makes the sound of a **j**.

Follow the routine next: Body motion, sentence, visualization, and writing.

Write PAGE on your whiteboard and add its family: age, sage, stage, cage, wage, rage, gage.

Children will make up a sentence they like, and then write and illustrate it. Ex: "At our age, sage is all the rage."

List G, Level 2

real

Show the children the SnapWord® card for REAL. Study the picture together and ask the children to comment on what they see. The picture for REAL is interesting. There are two boys and two snakes. Apparently one snake is real and the boys can't tell which one! Notice the middle **ea** spelling. This is a long **e** spelling.

Follow the routine next: Body motion, sentence, visualization, and writing.

Write REAL on your whiteboard and add its family: deal, heal, meal, peal, seal, teal, veal, zeal, squeal, steal.

Children will make up a sentence they like, and then write and illustrate it. Ex: "This meal is the real deal!" Or "The teal seal will squeal for this meal!"

side

Show the children the SnapWord® card for SIDE. Study the picture together and ask the children to comment on what they see. Notice that in this picture, there are four children - one for each letter in the word. Two kids are on one side, while the other two are on the other side. The sound spelling focus is of course the long **i** sound with Pinchy **e** on the end.

Follow the routine next: Body motion, sentence, visualization, and writing.

Write SIDE on your whiteboard and add its family: bide, hide, ride, tide, wide, bride, chide, glide, pride, slide, snide, stride.

Children will make up a sentence they like, and then write and illustrate it. Ex: "I like to hide by the side of the slide." Or "I will hide from the wide ride!"

size

Show the children the SnapWord® card for SIZE. Study the picture together and ask the children to comment on what they see. It looks as though the tall boy is trying to find a shirt that is the right size for his friend. There are two shirts on the floor, neither of which is going to work. Pinchy **e** is frowning - not really happy with how things are going. The focus is on sound spelling long **i** with Pinchy **e**.

Follow the routine next: Body motion, sentence, visualization, and writing.

Write SIZE on your whiteboard and add its family: prize, realize, resize, utilize, vaporize. Note the words are more advanced than usual, but if you cover IZE with your finger, the root words aren't so bad!

Children will make up a sentence they like, and then write and illustrate it. Ex: "I realize the size of the prize is huge!"

List G, Level 2

Show the children the SnapWord® card for SURE. Study the picture together and ask the children to comment on what they see. In this picture, the first girl, the **s**, is stirring sugar into her tea. The other child is the **e** which doesn't make any sound at all. The **ur** in the middle makes the sound of **er**. There are two sound spelling ideas in this word: One is that the **s** sounds like **sh**, and the other is the **ur** which sounds like **er**.

Follow the routine next: Body motion, sentence, visualization, and writing.

Write SURE on your whiteboard and in two columns, write words that go with the **s** or **ss** /SH/ and in the other column, the words that go with UR /ER/. S or SS words: sure, sugar, insure, assure, surely, issue, tissue. UR words: hurt, turn, fur, curl, surf, churn, burst, turkey, return, curt, urn, turf, spur, curb, cur.

Children will make up a sentence they like, and then write and illustrate it. Ex: "I am sure I want less sugar!" Or "It won't hurt to order surf and turf!"

List G, Level 3

above

Show the children the SnapWord® card for ABOVE. Study the picture together and ask the children to comment on what they see. The letters in this word are all floating lightly above the ground. There are a couple of interesting things about ABOVE. The first sound we hear is the schwa - /UH/ but it is spelled **a**. The other thing to note is that the **ove** sounds like **uv** rather than the long **o** sound as in "cove." The spelling pattern, therefore, is **o-e** that sounds like **uh** with a consonant in the place where the dash is.

Follow the routine next: Body motion, sentence, visualization, and writing.

Write ABOVE on your whiteboard and add its family: love, dove, shove, done, glove, some, come.

Children will make up a sentence they like, and then write and illustrate it. Ex: "I see the dove I love above us." Or "I'll shove my hand in the glove when I'm done."

among

Show the children the SnapWord® card for AMONG. Study the picture together and ask the children to comment on what they see. AMONG means "surrounded by something." In this picture, a huge bowl of beans is tucked among the people who have gathered to share food. You can share something among members of a group. The three letters in AMONG that have closed circles (**a**, **o**, and **g**) are the colorful people. The ending of the word **ng** is like the ending in words that end in **ing**, **ong**, and so forth, but this time the sound of the **ong** is /UNG/ rather than /ONG/ as in SONG. The sound of the **o** in AMONG is the same as the sound of the **o** in ABOVE.

Follow the routine next: Body motion, sentence, visualization, and writing.

Write AMONG on your whiteboard. The words which are spelled similarly (but don't SOUND the same) are: long, song, strong, gong, tong, prong, thong, wrong. So when the children are writing the words, don't include the O in the spelling. Just underline or highlight the final NG.

Children will make up a sentence they like, and then write and illustrate it. Ex: "Both the tong and the gong were strong!"

below

Show the children the SnapWord® card for BELOW. Study the picture together and ask the children to comment on what they see. I like the picture for BELOW because it looks like a fun thing to do - to make a play area below a table. Notice that the kids are made from the two tall letters in the words. The focus sound spelling is the final **ow** that sounds like long **o**.

Follow the routine next: Body motion, sentence, visualization, and writing.

Write BELOW on your whiteboard and add its family: bow, know, low, mow, row, sow, tow, blow, crow, flow, glow, grow, show, slow, snow, stow.

Children will make up a sentence they like, and then write and illustrate it. Ex: "We will throw the snow below the low glow of the moon." Or "I will mow the row that is slow to grow."

List G, Level 3

front

Show the children the SnapWord® card for FRONT. Study the picture together and ask the children to comment on what they see. This picture clearly shows what FRONT means! The boy in the front of the line looks super happy to be in that position. The other kids, not so much. Notice that the kid in the middle is an **o** who is not really making a short **o** sound. He is also saying "UH" like the **o** does in ABOVE and AMONG. Find little words inside of FRONT. There is ON, RON.

Follow the routine next: Body motion, sentence, visualization, and writing.

Write FRONT on your whiteboard. Now, here are some words that rhyme with FRONT but don't match its spelling: blunt, brunt, grunt, hunt, punt, shunt, stunt. Rhyming words: affront, confront. Compound words: forefront, lakefront, oceanfront, waterfront, seafront, shirtfront, shorefront, upfront.

Children will make up a sentence they like, and then write and illustrate it. Ex: "I'm upfront about liking seafront, lakefront, and oceanfront homes."

great

Show the children the SnapWord® card for GREAT. Study the picture together, noticing every little detail, and ask the children to comment on what they see. In this picture, the long **a** sound is spelled **ea**. We have the **a** in red in order to focus on the sound it makes. The **e** is there to help the **a** make a long sound, but he's not making his own sound at all. The guy in the picture looks super happy! I think he's yelling, "GREAT!" don't you?

Follow the routine next: Body motion, sentence, visualization, and writing.

Write GREAT on your whiteboard and add: break, steak.

Children will make up a sentence they like, and then write and illustrate it. Ex: "I'll take a break and have a great steak."

heavy

Show the children the SnapWord® card for HEAVY. Study the picture together and ask the children to comment on what they see. This picture is fun because the **h** and the **y** are lugging the other letters in the word and apparently they are HEAVY! The sound spelling focus for HEAVY is the **ea** that sounds like short **e**.

Follow the routine next: Body motion, sentence, visualization, and writing.

Write HEAVY on your whiteboard and add these words in which **ea** sounds like short **e**: bread, sweat, breakfast, meant, ready, already, spread, sweater, ahead, instead.

Children will make up a sentence they like, and then write and illustrate it. Ex: "I meant to already have bread spread with jam for breakfast!"

List G, Level 3

large

Show the children the SnapWord® card for LARGE. Study the picture together and ask the children to comment on what they see. In this picture we see Sarge who is Large! We have two focus sound spellings. The first is the AR (Bossy R) spelling and the other is the **ge** ending that sounds like **j**. This word only has three sounds in it. **l - r - j**.

Follow the routine next: Body motion, sentence, visualization, and writing.

Write LARGE on your whiteboard and then add two columns of words. Column 1 is for AR words: car, far, bar, tar, spar, star, start, park, part, mart, smart. The second column is for GE words: age, wage, cage, page, gage, sage, stage.

Children will make up a sentence they like, and then write and illustrate it. Ex: "The part of the park I like is far away." Or "On this page I see a cage on the stage."

often

Show the children the SnapWord® card for OFTEN. Study the picture together and ask the children to comment on what they see. The first clue in this picture is the hat the guy is wearing. It is the kind of hat that some artists like to wear. We find from the sentence on the back of the card that this artist is a potter and in the picture are some of the pots he OFTEN likes to make. Pay attention to which letters are the pots and which letter is the potter. In this word, we don't usually pronounce the **t**. It sounds like "offen." If your kids pronounce the **t** that is fine also!

Follow the routine next: Body motion, sentence, visualization, and writing.

Write OFTEN on your whiteboard and add its family: hen, den, men, pen, ten, glen, then, when, wren.

Children will make up a sentence they like, and then write and illustrate it. Ex: "When I got there, ten men put ten hens into the pen."

quick

Show the children the SnapWord® card for QUICK. Study the picture together and ask the children to comment on what they see. The picture for QUICK looks to me like what it says. The boy is quick to respond. Quick to jump. Quick to click his heels. You can see the other letters getting jostled because of his quick movements! The simple sound spellings in this word are **qu** that sounds like /KW/ and **ck** that sounds like **k**.

Follow the routine next: Body motion, sentence, visualization, and writing.

Write QUICK on your whiteboard and add its family: stick, wick, lick, kick, pick, sick, tick, brick, chick, click, flick, slick, thick, trick.

Children will make up a sentence they like, and then write and illustrate it. Ex: "I will pick a thick stick to kick."

List G, Level 3

ready

Show the children the SnapWord® card for READY. Study the picture together and ask the children to comment on what they see. This picture shows clearly the letter sounds we can hear. The towel is draped over the first syllable in the word: **re** (short **e**). Next we hear the boy and sound of **d**, and finally we hear the **y** sounding like long **e**. The focus sound spelling is **ea** sounding like short **e** like we saw in the lesson on HEAVY (see page 65).

Follow the routine next: Body motion, sentence, visualization, and writing.

Write READY on your whiteboard and add these words in which **ea** sounds like short **e**: bread, sweat, breakfast, meant, ready, already, spread, sweater, ahead, instead.

Children will make up a sentence they like, and then write and illustrate it. Ex: "I spread out the heavy, wet sweater."

really

Show the children the SnapWord® card for REALLY. Study the picture together and ask the children to comment on what they see. This word shows a child who is REALLY tall! In this word, the tricky sound spelling is the **ea** that (depending on who is speaking) might sound like short **i** ("rilly") or it might be long **e**.

Follow the routine next: Body motion, sentence, visualization, and writing.

Write REALLY on your whiteboard and add related words: real, seal, flea, meal, steal, mealy, veal, zeal, squeal, repeal, reveal, surreal.

Children will make up a sentence they like, and then write and illustrate it. Ex: "I really don't like a meal that is made of veal with fleas."

since

Show the children the SnapWord® card for SINCE. Study the picture together and ask the children to comment on what they see. In this funny picture, the **i** is a boy who has not combed his hair SINCE last July. Sitting on top of the **ce** which is sounding like **s**, is his unused comb! It has a whole string of zzzzzz's coming out of it meaning it has been sleeping from lack of use! The sound spelling focus for SINCE will be the ending that sounds like **s** but is spelled **ce**.

Follow the routine next: Body motion, sentence, visualization, and writing.

Write SINCE on your whiteboard and add its family: ice, nice, lice, mice, twice, splice, thrice, dice, rice, vice, price, slice, spice.

Children will make up a sentence they like, and then write and illustrate it. Ex: "Twice, since Monday, I saw nice mice on the ice."

List G, Level 3

sound

Show the children the SnapWord® card for SOUND. Study the picture together and ask the children to comment on what they see. This picture shows a boy listening to the SOUND of a train as it comes through town. Notice that the focus sound spelling (**ou**) is on the red of the engine. You can hear the **n** and **d** at the end of the word. Of course the listening boy is the initial **s**.

Follow the routine next: Body motion, sentence, visualization, and writing.

Write SOUND on your whiteboard and add its family: found, pound, mound, hound, wound, out, about, our, house, around, ground, astound, shout, stout, mouse.

Children will make up a sentence they like, and then write and illustrate it. Ex: "Give a shout when you've found the mouse and the hound going around the house."-

List G, Level 4

against

Show the children the SnapWord® card for AGAINST. Study the picture together and ask the children to comment on what they see. This picture shows a tug of war with girls against the boys. It is interesting that the girls are the round letters (**aga**), while the boys are not. The middle letter looks like the referee! Maybe she is saying, "Ready, get set, go!" The sound spelling for this lesson is **ai** that sounds like short **e**. The ending **nst** can be sounded out. Find the smaller words inside AGAINST. " A", "IN", "GAIN", "AGAIN".

Follow the routine next: Body motion, sentence, visualization, and writing.

Write AGAINST on your whiteboard and add its itty bitty family: said, again, (some might pronounce "captain" and "mountain" similiarly to "against" but to me the AI in these words sounds more like short i - /mountin/ or /captin/)

Children will make up a sentence they like, and then write and illustrate it. Ex: "I said again, 'I am against snakes!'"

almost

Show the children the SnapWord® card for ALMOST. Study the picture together and ask the children to comment on what they see. A group of climbers or hikers are going up the side of a steep mountain. They are almost to the top! Notice that the vowels are both round and colored. ALMOST is similar to two words pushed together: ALL and MOST - except there is only one l in ALL. The focus sound spelling today is a single **o** that says the long **o** sound.

Follow the routine next: Body motion, sentence, visualization, and writing.

Write ALMOST on your whiteboard and add some long O words: host, most, post, ghost, told, hold, gold, cold, old, bold, sold, also, don't, both.

Children will make up a sentence they like, and then write and illustrate it. Ex: "I almost sold most of the old gold."

already

Show the children the SnapWord® card for ALREADY. Study the picture together and ask the children to comment on what they see. The two people in this picture are clearing the table and said, "We've already cleared the table." But I see the silverware on the floor, so while the have cleared the table, they still have a bit of work to do! This word has the **al** beginning like in ALMOST. The sound spelling to focus on is the **ea** that says short **e**. We have studied this sound spelling before, so you may just review lessons on: heavy, ready.

Follow the routine next: Body motion, sentence, visualization, and writing.

Write ALREADY on your whiteboard and add words: bread, sweat, breakfast, meant, ready, spread, sweater, ahead, instead.

Children will make up a sentence they like, and then write and illustrate it. Ex: "I already spread the bread for breakfast."

List G, Level 4

behind

Show the children the SnapWord® card for BEHIND. Study the picture together and ask the children to comment on what they see. Obviously the word is behind the tree. Notice that to the left of the tree is the word BE. The **h** is behind the tree trunk. Then you can see IN to the right of the tree, then the **d** is a tall person who is listening to the **i** talk. Actually, **i** thinks **d** can't see him because he thinks he is behind the tree, but hmmm. It is not working is it? The sound spelling in this word is the **i** which sounds like long **i**.

Follow the routine next: Body motion, sentence, visualization, and writing.

Write BEHIND on your whiteboard and add its family: hind, mind, bind, kind, find, rind, wind (long I), blind, grind.

Children will make up a sentence they like, and then write and illustrate it. Ex: "In my mind I find I don't mind if you are kind."

certain

Show the children the SnapWord® card for CERTAIN. Study the picture together and ask the children to comment on what they see. In this picture are two boys. One is checking out the gate on the cage. To me it looks as if an animal escaped from the cage. The first half of the word - **cer** - is in front of the cage. It sounds like SIR. Next you see and hear the **t**. Then you hear **in** but it is spelled **ain**. In this word are three sound spellings. **C** in front of **E** that sounds like **s**, **er** or Bossy **r**, and finally the **ai** spelling that sounds like short **i**. The word can be split into sounds like this: C - ER - T - AI - N. Sounds like S-R-T-I-N.

Follow the routine next: Body motion, sentence, visualization, and writing.

Write CERTAIN on your whiteboard and add soft C words: cent, circle, center, ceiling. Add ER words: her, ever, serve, were. Finally, add AI /i/ words: certainly, mountain, captain.

Children will make up a sentence they like, and then write and illustrate it. Ex: "I'm certain the captain is on the mountain!" "The cent is in a circle in the center of the ceiling." "We were to serve her."

during

Show the children the SnapWord® card for DURING. Study the picture together and ask the children to comment on what they see. In this picture, parents are very patiently listening while their son talks non stop all during dinner. This picture highlights the **ur** /ER/ sound spelling by having two pink glasses on top of the **u** and the **r**. A pink cup also highlights the **ing** at the end of the word. Our focus sound spelling is going to be **ur** which sounds like **er**.

Follow the routine next: Body motion, sentence, visualization, and writing.

Write DURING on your whiteboard and add its family: hurt, turn, curl, burn, fur, surf, churn, return, turkey, burst.

Children will make up a sentence they like, and then write and illustrate it. Ex: "The surf will churn." "Return the turkey or I'll burst!" "It won't hurt to turn his fur in a curl."

List G, Level 4

either

Show the children the SnapWord® card for EITHER. Study the picture together and ask the children to comment on what they see. In this picture there are two options of color for the **e**'s. The child is saying, "Either color is fine with me." The colored arrows point to the two **e**'s in the word. This word has three sounds: E - TH - R spelled EI - TH - ER.

Follow the routine next: Body motion, sentence, visualization, and writing.

Write EITHER on your whiteboard and add its EI family: seize, ceiling, weird, leisure, receive, receipt, conceit, deceive, deceit, perceive, conceive.

Children will make up a sentence they like, and then write and illustrate it. Ex: "He received the receipt with conceit."

finally

Show the children the SnapWord® card for FINALLY. Study the picture together and ask the children to comment on what they see. In this picture we see a mom and her child out in the yard by a huge pile of fall leaves. It appears that they have been raking. The child is exclaiming, "We FINALLY finished raking!" The word has a common letter pattern for an ending. There are many words that end with ALLY. And the beginning is FIN but with a long **i** sound. Two small words inside of FINALLY are FIN and ALL. Push them together and add a final **y** and you have FINALLY!

Follow the routine next: Body motion, sentence, visualization, and writing.

Write FINALLY on your whiteboard and add: ally, dually, orally, royally, totally, usually, vitally, regally, globally. These words all have a root to which is added "ally." Ex: total turns into totally. Oral turns into orally. The first is a noun or adjective, and by adding ally, you are making the word into an adverb that tells how someone did something.

Children will make up a sentence they like, and then write and illustrate it. Ex: "Finally she was totally my ally!"

inside

Show the children the SnapWord® card for INSIDE. Study the picture together and ask the children to comment on what they see. This picture is great. It shows IN inside the house, and SIDE outside of the house! Because it is raining, we can be pretty sure the girl would love to be inside the house! The sound spelling for this word is primarily just long **i** spelled **i-e**. The word is also compound.

Follow the routine next: Body motion, sentence, visualization, and writing.

Write INSIDE on your whiteboard. Other words that go with INSIDE include: outside, beside, reside, upside, bedside, dayside, farside, offside, seaside, subside, topside, wayside.

Children will make up a sentence they like, and then write and illustrate it. Ex: "I reside outside by the seaside."

List G, Level 4

scared

Show the children the SnapWord® card for SCARED. Study the picture together and ask the children to comment on what they see. Oh my. In this picture the boy got scared when he heard an owl in the night. This word is interesting because there are so many little words in it: SCAR, CAR, ARE, CARE, RED, etc. The primary sound spelling is the **ar** that sounds like AIR.

Follow the routine next: Body motion, sentence, visualization, and writing.

Write SCARED on your whiteboard and add its family: fare, hare, bare, care, share, rare, spare, stare.

Children will make up a sentence they like, and then write and illustrate it. Ex: "The hare will care that it's bare!" "It's rare for the hare to share."

simple

Show the children the SnapWord® card for SIMPLE. Study the picture together and ask the children to comment on what they see. In this picture, a couple is having a very simple dinner: One humongous carrot! The lady highlights the **le** ending on the word. The table is made from the **mp** in the middle. The man is the **i**. The sound spelling is going to be the final **le** that just sounds like **l**.

Follow the routine next: Body motion, sentence, visualization, and writing.

Write SIMPLE on your whiteboard and add: people, purple, maple, little, sample, apple.

Children will make up a sentence they like, and then write and illustrate it. Ex: "The little people liked to sample a simple purple apple."

they're

Show the children the SnapWord® card for THEY'RE. Study the picture together and ask the children to comment on what they see. This picture shows that THEY'RE is made of two words: THEY and ARE. The black boot came in and kicked out the a from the ARE and now all that is left is THEY and RE. The concept for this word is contractions, but the sound spelling would be the **ey** that sounds like long **a**.

Follow the routine next: Body motion, sentence, visualization, and writing.

Write THEY'RE on your whiteboard. The other words that have EY /long A/ include prey, they, fey, grey, hey, whey, convey, heyday, obey, purvey, survey.

Children will make up a sentence they like, and then write and illustrate it. Ex: "Hey! When I survey, I see the grey prey."

List G, Level 5

although

Show the children the SnapWord® card for ALTHOUGH. Study the picture together and ask the children to comment on what they see. This picture does the job of breaking up a very long, potentially confusing word. The people on the ground say AL. The ones on the steps say TH, while those in the house just say OH together. This is another word that begins with ALL but with only one L. Other words that follow this same pattern include "already" and "almost." So AL is one spelling pattern and OUGH is another.

Follow the routine next: Body motion, sentence, visualization, and writing.

Write ALTHOUGH on your whiteboard. Then add a column for AL words: always, almond, alms, also, alright, almost, already. Now add the OUGH word when OUGH sounds like long O: though. Yep, that's it! There are other words that have the OUGH spelling: OUGH /off/ - cough. OUGH /oo/ - through. OUGH /ow/ - bough, slough (a bog), drought. OUGH /o/ - ought, thought, brought, bought, fought. OUGH /uf/ - rough, tough, enough, slough ("sluf" - as in to shed skin).

Children will make up a sentence they like, and then write and illustrate it. Ex: "It is true, though. Although I left for home, I never got there!"

beautiful

Show the children the SnapWord® card for BEAUTIFUL. Study the picture together and ask the children to comment on what they see. The picture for BEAUTIFUL is designed to help with reading and spelling the word! It starts with BE with the **a** grayed out because you don't hear it. Next is a blue **u**, then **ti**, then **ful**. You could have the children pronounce the word BE • A • U • TI • FUL. The only real sound spelling on which to focus is the FUL at the end.

Follow the routine next: Body motion, sentence, visualization, and writing.

Write BEAUTIFUL on your whiteboard. Add: fitful, eyeful, joyful, sinful, awful, lawful, artful, capful, lapful, useful.

Children will make up a sentence they like, and then write and illustrate it. Ex: "I got an eyeful of beautiful, artful flowers."

between

Show the children the SnapWord® card for BETWEEN. Study the picture together and ask the children to comment on what they see. This picture is super cute. The two boys are definitely BETWEEN three **e**'s. The purple rug helps highlight the kids who are BETWEEN the golden **e**'s. You can easily spot the small word BE at the beginning of the word. The sound spelling focus will be **ee** as long **e**.

Follow the routine next: Body motion, sentence, visualization, and writing.

Write BETWEEN on your whiteboard. Then add: teen, green, sheep, weep, wheels, cheese, tree, etc.

Children will make up a sentence they like, and then write and illustrate it. Ex: "The sheep will weep between the trees." Or "The sheep will ride on wheels of green cheese."

List G, Level 5

brother

Show the children the SnapWord® card for BROTHER. Study the picture together and ask the children to comment on what they see. In this picture, there are three brothers. The first is a fisherman, the second is talking and pointing, and the last one is getting ready to play soccer. In this word, the primary spelling pattern is the **O** that sounds like /uh/. There is also a digraph **th** and a final **er** (Bossy **r**).

Follow the routine next: Body motion, sentence, visualization, and writing.

Write "BROTHER on your whiteboard. Then add: other, mother, done, dove, love, shove, above.

Children will make up a sentence they like, and then write and illustrate it. Ex: "My mother saw my brother with the dove I love."

favorite

Show the children the SnapWord® card for FAVORITE. Study the picture together and ask the children to comment on what they see. In this picture we see some of the FAVORITE things of a child. This word is easy to spell if you first write FAVOR and then add ITE. The spelling pattern is **or** that sounds like **er**.

Follow the routine next: Body motion, sentence, visualization, and writing.

Write FAVORITE on your whiteboard. Add other /OR/ words: favor, worth, worm, word, world, vigor, armor, worse, work.

Children will make up a sentence they like, and then write and illustrate it. Ex: "My favorite worm in the world will work with vigor."

important

Show the children the SnapWord® card for IMPORTANT. Study the picture together and ask the children to comment on what they see. This picture breaks up the long word IMPORTANT into chunks. There is **im** (the boy being impressed) and **p** (the important man), a girl who is **or**, and another girl who is a **t** and then of course there is **ant**. None of the sounds are difficult, so help kids focus in on what they are hearing when they say each syllable, and guide them to write what they hear. IM-POR-TANT. The only sound spelling in this word is **or**, but even this chunk sounds like it looks.

Follow the routine next: Body motion, sentence, visualization, and writing.

Write IMPORTANT on your whiteboard. Other OR words include: for, porch, storm, reform, deform, port, sport, stupor, actor, labor, vapor, etc.

Children will make up a sentence they like, and then write and illustrate it. Ex: "The important actor will labor on the porch during the storm."

List G, Level 5

instead

Show the children the SnapWord® card for INSTEAD. Study the picture together and ask the children to comment on what they see. The picture shows a boy (**in**) with big muscles who is saying we should eat broccoli instead of donuts. The **ea** is the table holding the broccoli. The sound spelling is one we've already visited in this manual. It is **ea** saying short **e**. You can hear all the sounds otherwise.

Follow the routine next: Body motion, sentence, visualization, and writing.

Write INSTEAD on your whiteboard. Let's add other ea words: bread, sweat, breakfast, meant, ready, spread, sweater, ahead.

Children will make up a sentence they like, and then write and illustrate it. Ex: "I meant to have bread instead of this for breakfast!"

perhaps

Show the children the SnapWord® card for PERHAPS. Study the picture together and ask the children to comment on what they see. Oh my. There's an elephant in the room. Her name is Babs. PERHAPS has two syllables that make it easy to spell and read. PER and HAPS. **er** is the only sound spelling to be concerned about.

Follow the routine next: Body motion, sentence, visualization, and writing.

Write PERHAPS on your whiteboard. Other ER words can include: berm, germ, term, fern, tern, stern, her, doer, over, rover, mover.

Children will make up a sentence they like, and then write and illustrate it. Ex: "Perhaps the mover will take her over the berm."

possible

Show the children the SnapWord® card for POSSIBLE. Study the picture together and ask the children to comment on what they see. Here's Babs brother Bob! It is possible the chair will not hold him! Again, break the word into syllables. POSS - I - BLE. The sound spelling is the **le** ending that sounds like **l**.

Follow the routine next: Body motion, sentence, visualization, and writing.

Write POSSIBLE on your whiteboard. Let's add other LE words: maple, little, purple, sample, apple, simple, etc.

Children will make up a sentence they like, and then write and illustrate it. Ex: "It's possible they love the simple maple apples.

List G, Level 5

probably

Show the children the SnapWord® card for PROBABLY. Study the picture together and ask the children to comment on what they see. This is elephant number three! His name is Bab and the child in the picture is saying that Bab probably won't fit through the red door. With this word, the image helps with decoding the word. **P** is a child, **o** is the red door, bab is colored green to help it stand out. (Also it is the name of the elephant!) Finally the word ends with "**ly**". The primary sound spelling in this word is the final **ly**. All the other sounds are single letter sounds that can be sounded out.

Follow the routine next: Body motion, sentence, visualization, and writing.

Write PROBABLY on your whiteboard. Let's add other words: barely, loosely, directly, scarcely, certainly, possibly.

Children will make up a sentence they like, and then write and illustrate it. Ex: "We probably will barely get back in time."

several

Show the children the SnapWord® card for SEVERAL. Study the picture together and ask the children to comment on what they see. In this picture several kids are looking for a fish, several are deciding if they want to fish or not. If you look for smaller words inside SEVERAL you can find EVE, EVER, SEVER, and **al**, possibly the name of the tallest boy. There are two sound spellings in this word: **er** and ending **al**. Notice that while the AL ending has two letters, you can only hear the simple sound **l** makes. SEVERAL actually sounds like S-E-V-R-L.

Follow the routine next: Body motion, sentence, visualization, and writing.

Write SEVERAL on your whiteboard separating the word into syllables. SE-VER-AL. Brainstorm some ER words: ever, never, lever, sever, her. Now words with the AL ending: rural, sabal, natural, general, loyal, royal, final, normal, rural, festival.

Next children will make up a sentence they like, and then write and illustrate it. Ex: "Several were loyal to the rural royal."

themselves

Show the children the SnapWord® card for THEMSELVES. Study the picture together and ask the children to comment on what they see. This huge word is actually not that bad if we separate it into two words, THEM and SELVES. The picture helps by doing the separation for us. The only tricky part of this word is the fact that the final **es** sounds like a **z**.

Follow the routine next: Body motion, sentence, visualization, and writing.

Write THEMSELVES on your whiteboard. Let's brainstorm some words that end with ES that sounds like Z. Goes, does, potatoes, tomatoes, volcanoes, thieves, knives, shelves.

Next children will make up a sentence they like, and then write and illustrate it. Ex: "They will keep the potatoes, tomatoes, and knives on the shelves for themselves."

SNAPWORDS LIST N1

List N1 Words - ABC order:	List N1 Words by level:
ball bird book boy chair children city clothes cloud country crab desk ears eyes fire flower food friend giant girl grass hair head house insect island lizard ocean paper people planet plant rain river rock sand school shirt shoe sign snake snow spider stake stick storm street sun table teacher town tree water wind woman words world worm year	**Level 1:** food, school, ball, grass, house, fire, river, wind, giant, ears, city, island **Level 2:** tree, sun, water, friend, boy, girl, shoe, ocean, table, woman, people, cloud **Level 3:** town, country, storm, rain, planet, crab, head, lizard, snake, spider, stick **Level 4:** worm, bird, words, plant, shirt, flower, paper, insect, book, teacher, desk, sign **Level 5:** stake, eyes, hair, chair, world, clothes, snow, children, year, street, rock, sand

How To Teach   List N1, Level 1

food

Show the children the SnapWord® card for FOOD. Study the picture together and ask the children to comment on what they see. Your students will have already studied the long **oo** sound spelling, so FOOD should be a breeze. Just note the **f** banana, the **o** pizza and peach, and the **d** delighted girl! She must love bananas, pizzas, and peaches!

Follow the routine next: Body motion, sentence, visualization, and writing.

Write FOOD on your whiteboard. Add: "moon, boon, mood, brood, boom, loom, doom, room, zoon, pool, tool, fool, cool."

Children will make up a sentence they like, and then write and illustrate it. Ex: "We ate our cool food under the moon by the pool."

school

Show the children the SnapWord® card for SCHOOL. Study the picture together and ask the children to comment on what they see. The great thing about this picture is that the one letter that you cannot hear (**h**) is highlighted with the flag that flies over the entrance to the school. The shape of the building also accommodates the other tall letter (**l**). Of course the children know the **oo** spelling and can hear the **s** and the **c**.

Follow the routine next: Body motion, sentence, visualization, and writing.

Write SCHOOL on your whiteboard and refer back to the word list for FOOD.

Children will make up a sentence they like, and then write and illustrate it. Ex: "We have a cool room at school!"

ball

Show the children the SnapWord® card for BALL. Study the picture together and ask the children to comment on what they see. This is one word that might be a no-brainer for your children, so don't belabor the point. If it is already a familiar word, do visual imprinting, make sure they can easily use it in writing sentences, including spelling it correctly, and move on to the next lesson.

Write BALL on your whiteboard and add other LL words as desired or needed. "call, fall, gall, hall, mall, pall, tall, wall, small, squall, stall.

Sample sentences: "In the mall we play ball in the hall by the wall." "The small ball is in the stall down the hall in the mall."

List N1, Level 1

grass

Show the children the SnapWord® card for GRASS. Study the picture together and ask the children to comment on what they see. There is nothing really tricky about GRASS. The only thing your child(ren) will not be able to hear is the double **ss**, but this is where visual imprinting comes in handy.

Follow the routine next: Body motion, sentence, visualization, and writing.

Write GRASS on your whiteboard and add: "bass, lass, mass, pass, brass, class, glass."

Children will make up a sentence they like, and then write and illustrate it. Ex: "We will pass the class sitting in the grass."

house

Show the children the SnapWord® card for HOUSE. Study the picture together and ask the children to comment on what they see. In this picture, the focus is on the **h** which is actually the house! Comment on the **ou** spelling which says OW! And the **se** spelling which says the "ssss" sound at the end of house.

Follow the routine next: Body motion, sentence, visualization, and writing.

Write HOUSE on your whiteboard and add: "douse, mouse, rouse, souse, blouse, grouse, spouse."

Children will make up a sentence they like, and then write and illustrate it. Ex: "We have a mouse in the house!" "The mouse is in the house, but not the grouse!"

###

Show the children the SnapWord® card for FIRE. Study the picture together and ask the children to comment on what they see. Oh no! In this picture, the **r** looks like it is burning up! In the meantime, the **i** and the **e** are leaning away from the burning **r**. In spite of the fire, the **e** is finding time to reach around and pinch **i** so it will say its name! Imagine that!

Follow the routine next: Body motion, sentence, visualization, and writing.

Write FIRE on your whiteboard and add: "hire, tire, wire, spire."

Children will make up a sentence they like, and then write and illustrate it. Ex: "Do put the tire and the wire in the fire!" "The spire is on fire!"

List N1, Level 1

river

Show the children the SnapWord® card for RIVER. Study the picture together and ask the children to comment on what they see. In this peaceful picture I can almost hear the sound of the water flowing. It is so peaceful, in fact, that the **e** is floating along, helping **r** make its sound rather than pinching the **i**. The first **r** is over a rock pile, while the last **r** looks like its surfing on a ripple of water.

Follow the routine next: Body motion, sentence, visualization, and writing.

Write RIVER on your whiteboard and add: "liver, quiver, giver, shiver, sliver, deliver, upriver."

Children will make up a sentence they like, and then write and illustrate it. Ex: "She will shiver at the liver in the river." "He will deliver the quiver to the river." "I got a sliver at the river."

wind

Show the children the SnapWord® card for WIND. Study the picture together and ask the children to comment on what they see. This word has no tricky spelling issues, so let the image deliver the spelling during visual imprinting. You can almost feel the wind in this picture as it blows the word to a slant, and blows the girl's hair and scarf back. In fact, everything in the picture is blowing in the wind towards the right. Find the small words in WIND: I, IN, and WIN. Of course the **d** is the girl.

Follow the routine next: Body motion, sentence, visualization, and writing.

Write WIND on your whiteboard. There are virtually no words that rhyme with WIND, but there are many words that share the IND spelling with a long I. These include: "bind, hind, find, kind, mind, rind, wind, blind, grind."

Children will make up a sentence they like, and then write and illustrate it. Ex: "The wind is wild." "I find I do not mind the wind."

giant

Show the children the SnapWord® card for GIANT. Study the picture together and ask the children to comment on what they see. The very large **i** is hard to miss and it also shows the meaning of the word GIANT. Notice also the ant parade behind the children. They draw attention to the small word ANT in GIANT. Another small word is AN. It is important to focus on **g** at the beginning of the word because the **g** is making the sound of a **j**. This guy is on his knees in awe of how very tall the giant is!

Follow the routine next: Body motion, sentence, visualization, and writing.

Write GIANT on your whiteboard and add: "grant, slant, enchant, implant, replant."

Children will make up a sentence they like, and then write and illustrate it. Ex: "The giant will enchant us as he can replant the trees." "We will replant the giant trees."

List N1, Level 1

ears

Show the children the SnapWord® card for EARS. Study the picture together and ask the children to comment on what they see. It is hard to miss the two ears which are the **e** and the final **s**. It is important to notice the **a** following the **e** as the two letters together make the long **e** sound.

Follow the routine next: Body motion, sentence, visualization, and writing.

Write EARS on your whiteboard and add: "fears, dears, hears, gears, spears, nears, tears, years."

Children will make up a sentence they like, and then write and illustrate it. Ex: "He hears with his ears" "She fears as the gears are near her ears."

city

Show the children the SnapWord® card for CITY. Study the picture together and ask the children to comment on what they see. In this word, we are in the city at night. The moon is peeking out at the top of the **c** which is making a soft **s** sound. **I** and **t** are tall buildings, while **y** and **t** have a wire holding a street light between them. There is also an interesting slanting building following the slant of the **y**.

Follow the routine next: Body motion, sentence, visualization, and writing.

Write CITY on your whiteboard and add: "pity, unity, amity, cavity, sanity."

Children will make up a sentence they like, and then write and illustrate it. Ex: "We want unity and amity in the city."

island

Show the children the SnapWord® card for ISLAND. Study the picture together and ask the children to comment on what they see. This picture clearly shows what an island is. There are even sharks swimming around the perimeter of the island to show there is water all the way around it. ISLAND looks like two words IS and LAND and this would be a great way to remember its spelling (The ISLAND IS LAND – not water.) The first tall letter, **i** is a palm tree, and the last tall letter, **d** looks like it is almost in the water. The **a** in the middle frames the sun which is shining just above the water.

Follow the routine next: Body motion, sentence, visualization, and writing.

Write ISLAND on your whiteboard and add: "band, stand, grand, land, expand, strand."

Children will make up a sentence they like, and then write and illustrate it. Ex: "We want to expand the land on the island." "The land on the island is small."

List N1, Level 2

tree

Show the children the SnapWord® card for TREE. Study the picture together and ask the children to comment on what they see. The **t** IS the tree in this picture. Notice the roots coming out of the bottom of the trunk and the worm crawling away. The long **e** spelling is **ee** and certainly the students are familiar with this sound spelling. But just in case, we will follow the routine!

Follow the routine next: Body motion, sentence, visualization, and writing.

Write TREE on your whiteboard and add: "bee, see, agree, flee, knee, free, three."

Children will make up a sentence they like, and then write and illustrate it. Ex: "We see a bee in the tree flying free." "I agree we see a bee by the tree."

sun

Show the children the SnapWord® card for SUN. Study the picture together and ask the children to comment on what they see. This picture is different because the color of the letters matches the sun rather than being black. The letters are also curved around the sun to show its round shape.

Follow the routine next: Body motion, sentence, visualization, and writing.

Write SUN on your whiteboard and add: "fun, bun, stun, shun, begun, rerun."

Children will make up a sentence they like, and then write and illustrate it. Ex: "A rerun on TV was about fun in the sun." "She had begun to shun the sun."

water

Show the children the SnapWord® card for WATER. Study the picture together and ask the children to comment on what they see. We are in the desert with a cowboy looking for water! Fortunately, the cowboy found a **w** full of blue water! The **t** is the cowboy and the **er** finish the word surrounded by cactus. The sound **a** makes is like the sound in "father, want, wash, what."

Follow the routine next: Body motion, sentence, visualization, and writing.

Write WATER on your whiteboard and add these great compound water words: "backwater, seawater, saltwater, tidewater, wastewater, whitewater, rainwater, breakwater, freshwater."

Children will make up a sentence they like, and then write and illustrate it. Ex: "Saltwater and freshwater are not the same." "Rainwater and wastewater are not the same."

List N1, Level 2

friend

Show the children the SnapWord® card for FRIEND. Study the picture together and ask the children to comment on what they see. **F** starts the word and he's just thrown the ball to his friend, tall letter **d**. **R** is huddled close to the **f** boy. We can see the little word END inside of FRIEND, but the one letter that is just, well, standing there is the **i**. Use the sentence "I am your friend to the end" to help children remember the **i**. Rely heavily on visual imprinting – noticing that the lines that show the path of the ball start just over the **i**.

Follow the routine next: Body motion, sentence, visualization, and writing.

Write FRIEND on your whiteboard and add: "bend, send, spend, lend, mend, tend, amend, attend."

Children will make up a sentence they like, and then write and illustrate it. Ex: "My friend and I will attend." "My friend will mend and then send me the ball that popped."

boy

Show the children the SnapWord® card for BOY. Study the picture together and ask the children to comment on what they see. Tall **b** is the boy, and boy is he busy. He is walking along while he listens to music, and he might be humming along. **Oy** work together to make the diphthong. The **o** is just floating between the **b** and the final **y**, both of which are touching the ground.

Follow the routine next: Body motion, sentence, visualization, and writing.

Write BOY on your whiteboard and add: "Joy, enjoy, toy, Troy, soy, ploy, employ, convoy."

Children will make up a sentence they like, and then write and illustrate it. Ex: "The boy with the toy is Troy." "Joy will enjoy the toy with Troy."

girl

Show the children the SnapWord® card for GIRL. Study the picture together and ask the children to comment on what they see. This is a great picture because the **i** is the girl and it is also the one letter we can't really hear. The **ir** makes the sound ER which is spelled five different ways (er, ur, ir, or, ear).

Follow the routine next: Body motion, sentence, visualization, and writing.

Write GIRL on your whiteboard and add: "whirl, swirl, twirl." Other words using IR that sounds like ER include: "dirt, bird, circus, fir, flirt, squirt, sir, stir, whir, gird, third, birth, firth, girth, mirth." When you have your students write these words, ask them to underline the IR spelling in each one or write the IR in a different color.

Children will make up a sentence they like, and then write and illustrate it. Ex: "Shirl is a girl who likes to swirl and twirl."

List N1, Level 2

shoe

Show the children the SnapWord® card for SHOE. Study the picture together and ask the children to comment on what they see. The **sh** in this picture is the heel of the shoe, while the **oe** combine to both make the **oo** sound and the toe of the shoe.

Follow the routine next: Body motion, sentence, visualization, and writing.

Write SHOE on your whiteboard and add: "canoe, overshoe, snowshoe, horseshoe."

Children will make up a sentence they like, and then write and illustrate it. Ex: "She took a toe shoe, a snowshoe, and a horseshoe into her canoe. What?!"

ocean

Show the children the SnapWord® card for OCEAN. Study the picture together and ask the children to comment on what they see. Note the eel slithering in and out of the **o** and the **c**. An anchor is behind the **a** but overlapping the **e** and the **n**. In this word, the **ce** combine to sound like SH.

Follow the routine next: Body motion, sentence, visualization, and writing.

Write OCEAN on your whiteboard and add: "licorice, crustaceous, herbaceous." Big words these! The first word is pronounced LI-CO-RISH. In crustaceous and herbaceous, the CE says SH and the OUS says US. Your more intrepid students will want to tackle these longer words, but if all don't want to, that is fine. The main point is to show that the CE spelling is out there used in larger words.

Children will make up a sentence they like, and then write and illustrate it. Ex: "There is no licorice in the ocean!"

table

Show the children the SnapWord® card for TABLE. Study the picture together and ask the children to comment on what they see. The first tall letter, **t**, is the boy standing at the table. The **a** is the stool with a cushion on it. Two tall letters, **b** and **l** make the table legs, and the final **e** which is working with the **l** to say the sound of **l** is the other stool. Rely heavily on visual imprinting with this one.

Follow the routine next: Body motion, sentence, visualization, and writing.

Write TABLE on your whiteboard and add: "able, Mable, stable, gable, sable, cable, enable, unable, disable."

Children will make up a sentence they like, and then write and illustrate it. Ex: "We are able to make the table stable." "Mable will disable the cable by the table."

List N1, Level 2

woman

Show the children the SnapWord® card for WOMAN. Study the picture together and ask the children to comment on what they see. In this picture, the **o** is the woman. Notice that the **m** and **n** each have a mountain behind them. There are three flowers bringing attention to the small word inside WOMAN: MAN. There is also A and AN.

Follow the routine next: Body motion, sentence, visualization, and writing.

Write WOMAN on your whiteboard and add: "plan, span, human, Roman, chairwoman, saleswoman, spokeswoman, catwoman."

Children will make up a sentence they like, and then write and illustrate it. Ex: "The woman is a Roman who plays Catwoman." "The woman has a plan."

people

Show the children the SnapWord® card for PEOPLE. Study the picture together and ask the children to comment on what they see. This picture looks fun! There are several people hanging around, possibly waiting for a bus or train. I recognize the WOMAN and the BOY – the **p** and the **e**. There is a gray **o** because we can't hear that sound (it is working with **e** to make the long **e** sound). Next, **p** and **l** are chatting, while the final **e** is quietly working with **l** to make his sound. This is how to break up PEOPLE: P-EO-P-LE. Each grouping makes one sound.

Follow the routine next: Body motion, sentence, visualization, and writing.

Write PEOPLE on your whiteboard and add: "maple, apple, triple, staple, simple, temple, ample, triple."

Children will make up a sentence they like, and then write and illustrate it. Ex: "The people all want a maple apple."

cloud

Show the children the SnapWord® card for CLOUD. Study the picture together and ask the children to comment on what they see. The word CLOUD is floating in a cloud in the sky, fading off on the right. Notice the sun under the **o** and the bird flying under the **u**. The final **d** is right over a tiny town.

Follow the routine next: Body motion, sentence, visualization, and writing.

Write CLOUD on your whiteboard and add: "aloud, loud, proud, shroud."

Children will make up a sentence they like, and then write and illustrate it. Ex: "The cloud will shroud the trees." "I am proud of our cloud. It is not loud."

List N1, Level 3

town

Show the children the SnapWord® card for TOWN. Study the picture together and ask the children to comment on what they see. In this picture, each letter except **w** is a building or house. The **t** is tall and lavender, the **o** is a little round house with smoke coming from the chimney, while the **n** is red with a green roof. Notice the tree to the left of the word and the stop sign to the right. Focus on the **ow** spelling.

Follow the routine next: Body motion, sentence, visualization, and writing.

Write TOWN on your whiteboard and add: "down, gown, brown, clown, frown, drown."

Children will make up a sentence they like, and then write and illustrate it. Ex: "In this town we have a clown in a brown gown." "We like to go down into town."

country

Show the children the SnapWord® card for COUNTRY. Study the picture together and ask the children to comment on what they see. Everything in this picture has to do with being in the country. Notice all the details. The snake is crawling through the **c** and the **o**. There is a red bird on the **n** while the **t** is a tree. The **r** is an old-fashioned water pump and the **y** contains a nest for the red birds. The sun is shining on the **ou** as they work together to make the UH sound in COUNTRY. Notice the short word TRY at the end of the word.

Follow the routine next: Body motion, sentence, visualization, and writing.

Write COUNTRY on your whiteboard and add: "pastry, poultry, pantry, wintry, poetry, entry."

Children will make up a sentence they like, and then write and illustrate it. Ex: "In the country we have pastry and poultry." "The country entry is wintry poetry."

storm

Show the children the SnapWord® card for STORM. Study the picture together and ask the children to comment on what they see. Looking at this picture, I can see and hear the storm! There is thunder and lightning and rain splashing into the puddle. Find the small word OR inside STORM. Notice that the lightning bolts are about to hit the **t** and the **r**.

Follow the routine next: Body motion, sentence, visualization, and writing.

Write STORM on your whiteboard and add: "dorm, form, norm, reform" and then add compound words: "hailstorm, rainstorm, thunderstorm, sandstorm, firestorm, snowstorm."

Children will make up a sentence they like, and then write and illustrate it. Ex: "The storm was the norm." "I like a snowstorm more than a firestorm."

List N1, Level 3

rain

Show the children the SnapWord® card for RAIN. Study the picture together and ask the children to comment on what they see. This picture is pretty funny because the cloud and all the rain are JUST over the girl with her yellow umbrella. Notice that the other letters have little evergreen trees by them and peek through them. Maybe the **i**, the girl, is complaining that the rain is following her around! Focus on long **a** spelled **ai**.

Follow the routine next: Body motion, sentence, visualization, and writing.

Write RAIN on your whiteboard and add: "train, brain, strain, sprain, gain, grain, chain, main, Spain."

Children will make up a sentence they like, and then write and illustrate it. Ex: "The rain in Spain falls mainly on the plain." "The grain in the train is in the rain."

planet

Show the children the SnapWord® card for PLANET. Study the picture together and ask the children to comment on what they see. This word is two syllables: PLA-NET. Notice that the a looks a bit like Saturn – the planet with a large ring around it. Let children share what they notice about the stars and their placement on the various letters.

Follow the routine next: Body motion, sentence, visualization, and writing.

Write PLANET on your whiteboard and add: "hornet, sonnet, bonnet, garnet, magnet."

Children will make up a sentence they like, and then write and illustrate it. Ex: "The sonnet was about the planet." "On our planet, the garnet was a magnet."

crab

Show the children the SnapWord® card for CRAB. Study the picture together and ask the children to comment on what they see. In this picture, it looks like the crab is trying to hug the word – notice his pinchers going around the **c** and the **b**. This is a relatively easy word, so follow the lead of the child(ren).

Follow the routine next: Body motion, sentence, visualization, and writing.

Write CRAB on your whiteboard and add: "drab, flab, grab, nab, lab, slab, stab."

Children will make up a sentence they like, and then write and illustrate it. Ex: "The crab is not drab!" "I will try and grab the crab."

List N1, Level 3

head

Show the children the SnapWord® card for HEAD. Study the picture together and ask the children to comment on what they see. The **h** and the **d** have ears in them while the **ea** each have an eye peeking out as they work together to make the short **e** sound.

Follow the routine next: Body motion, sentence, visualization, and writing.

Write HEAD on your whiteboard and add: "bread, sweat, dead, dread, read, thread, spread, tread, stead."

Children will make up a sentence they like, and then write and illustrate it. Ex: "It is in my head that I want bread." "I dread having sweat on my head."

lizard

Show the children the SnapWord® card for LIZARD. Study the picture together and ask the children to comment on what they see. This picture helps with the **ar** that sounds like **er**. Notice the legs coming out of the **a**.

Follow the routine next: Body motion, sentence, visualization, and writing.

Write LIZARD on your whiteboard and add: "blizzard, gizzard, wizard."

Children will make up a sentence they like, and then write and illustrate it. Ex: "In the blizzard, the lizard held his gizzard." "The wizard is a lizard."

snake

Show the children the SnapWord® card for SNAKE. Study the picture together and ask the children to comment on what they see. Notice how nicely the **s** fits into the curve of the yellow snake. The **k** is a twig with leaves, while the final, pinchy **e** is a boulder.

Follow the routine next: Body motion, sentence, visualization, and writing.

Write SNAKE on your whiteboard and add: "bake, stake, rake, brake, make, take, lake."

Children will make up a sentence they like, and then write and illustrate it. Ex: "I hit the brake when I saw the snake." "The snake was by the stake at the lake."

List N1, Level 3

spider

Show the children the SnapWord® card for SPIDER. Study the picture together and ask the children to comment on what they see. Which letter is the largest in SPIDER? (the **p**). Which letter is on the spider's head? (the **r**). Ask the children to comment about the placement of the letters on the spider.

Follow the routine next: Body motion, sentence, visualization, and writing.

Write SPIDER on your whiteboard and add: "cider, glider, rider, slider."

Children will make up a sentence they like, and then write and illustrate it. Ex: "The spider was the rider on the glider." "I saw a spider on my slider!"

stick

Show the children the SnapWord® card for STICK. Study the picture together and ask the children to comment on what they see. Cool picture! The **t** is the stick which the boy (the **i**) is holding. The **ck** work together to make the **k** sound at the end of the word. The rock draws attention to the fact they are working together.

Follow the routine next: Body motion, sentence, visualization, and writing.

Write STICK on your whiteboard and add: "brick, nick, lick, trick, wick."

Children will make up a sentence they like, and then write and illustrate it. Ex: "Nick did a trick with a stick." "The stick is slick!"

List N1, Level 4

worm

Show the children the SnapWord® card for WORM. Study the picture together and ask the children to comment on what they see. This simple picture does an amazing job of reminding the children that ER is spelled with an **o**. The worm is hard to forget, right? The leaf is placed neatly behind the **or** spelling.

Follow the routine next: Body motion, sentence, visualization, and writing.

Write WORM on your whiteboard and add: "work, world, worry, color, worst, flavor, sailor, labor, tractor, actor, honor."

Children will make up a sentence they like, and then write and illustrate it. Ex: "The worm had the worst flavor!" "I worry about the color of the worm!"

bird

Show the children the SnapWord® card for BIRD. Study the picture together and ask the children to comment on what they see. The position of the red bird behind the **ir** is on purpose to help children visually remember that the **ir** says **er**. Note also that the **b** and **d** have their bellies facing each other.

Follow the routine next: Body motion, sentence, visualization, and writing.

Write BIRD on your whiteboard and add: "sir, dirt, skirt, third, girl, bluebird redbird."

Children will make up a sentence they like, and then write and illustrate it. Ex: "The third girl saw the first bird in the dirt."

words

Show the children the SnapWord® card for WORDS. Study the picture together and ask the children to comment on what they see. Look at all those WORDS on the **o** and the **d**! In this word, the **or** says **er** just like in WORM.

Follow the routine next: Body motion, sentence, visualization, and writing.

Write WORDS on your whiteboard and reference the word list for WORM.

Children will make up a sentence they like, and then write and illustrate it. Ex: "The words on the word are of many colors." "The actor likes words."

List N1, Level 4

plant

Show the children the SnapWord® card for PLANT. Study the picture together and ask the children to comment on what they see. What a cool plant! Note the position of the flowers on the **p**, the tall **l**, and the **n**. Find these small words: A, AN, and ANT.

Follow the routine next: Body motion, sentence, visualization, and writing.

Write PLANT on your whiteboard and add: "ant, pant, chant, slant."

Children will make up a sentence they like, and then write and illustrate it. Ex: "The plant is on a slant." "The ant is on the plant."

shirt

Show the children the SnapWord® card for SHIRT. Study the picture together and ask the children to comment on what they see. I love this picture. The **sh** is highlighted by the red cap. The **i** is the first boy with the shirt that is way too small! This boy points out that **ir** sounds like **er**. The boy who is the final **t** has a shirt that is too large! What to do...what to do?

Follow the routine next: Body motion, sentence, visualization, and writing.

Write SHIRT on your whiteboard and use the word lists from GIRL and BIRD.

Children will make up a sentence they like, and then write and illustrate it. Ex: "Sir, your shirt is too big!" "My shirt has dirt on it!"

flower

Show the children the SnapWord® card for FLOWER. Study the picture together and ask the children to comment on what they see. This flower is stupendous! The girl is the **f** and the **ow** is working together to make the **ow** sound of course. Sometimes **ou** makes the sound **ow**, but in this picture, we see that the **w** is filled with potting soil. The **er** at the end of the word of course says **er**.

Follow the routine next: Body motion, sentence, visualization, and writing.

Write FLOWER on your whiteboard and add: "power, tower, cower, shower, bower."

Children will make up a sentence they like, and then write and illustrate it. Ex: "The flower grows up the tower." "The flower bower is on the tower."

List N1, Level 4

paper

Show the children the SnapWord® card for PAPER. Study the picture together and ask the children to comment on what they see. In this picture, the second **p** is a girl who loves to draw on paper! Note the final **er** is holding a stack of paper. Notice the books circling the girl's feet.

Follow the routine next: Body motion, sentence, visualization, and writing.

Write PAPER on your whiteboard and add: "taper, caper, shaper, draper, scraper."

Children will make up a sentence they like, and then write and illustrate it. Ex: "The girl is a shaper of paper!" "We had a paper caper!"

insect

Show the children the SnapWord® card for INSECT. Study the picture together and ask the children to comment on what they see. I love this picture. It shows the two syllables separated – **in** on the head, and **sect** on the body. Note that the letters **e** and **c** are very nearly the same shape.

Follow the routine next: Body motion, sentence, visualization, and writing.

Write INSECT on your whiteboard and add: "detect, inspect, respect, aspect, reflect, object."

Children will make up a sentence they like, and then write and illustrate it. Ex: "I will inspect to detect the insect." "I object with respect to an insect."

book

Show the children the SnapWord® card for BOOK. Study the picture together and ask the children to comment on what they see. In this picture we see the word neatly cut in half by the "ditch" in between the pages. The division comes right between the two **o**'s that work together to say OO.

Follow the routine next: Body motion, sentence, visualization, and writing.

Write BOOK on your whiteboard and add: "took, look, cook, brook, snook, good, wood."

Children will make up a sentence they like, and then write and illustrate it. Ex: "She took a good look at the book." "In the book I see a snook in the brook."

List N1, Level 4

teacher

Show the children the SnapWord® card for TEACHER. Study the picture together and ask the children to comment on what they see. In this picture, the initial letter is a **t** holding the flag. Then comes **ea** who work together to say the long **e** sound. The teacher is the **h** that works with the **c** to say CH. At the end the **er** appears.

Follow the routine next: Body motion, sentence, visualization, and writing.

Write TEACHER on your whiteboard and add: "preacher, creature, feature, bleacher."

Children will make up a sentence they like, and then write and illustrate it. Ex: "The teacher saw the feature in the bleacher."

desk

Show the children the SnapWord® card for DESK. Study the picture together and ask the children to comment on what they see. The **d** is the student preparing to work at his desk. Notice that the corner of the desk is resting on the **e**. The top edge of the desk rests on the **k** which also supports the lamp and pencil holder.

Follow the routine next: Body motion, sentence, visualization, and writing.

Write DESK on your whiteboard and add two compound words: "copydesk, desktop."

Children will make up a sentence they like, and then write and illustrate it. Ex: "I work at a desk." "My paper is on the desktop."

sign

Show the children the SnapWord® card for SIGN. Study the picture together and ask the children to comment on what they see. The images in this word are critically important because there is a **g** lurking in there that doesn't apparently say anything at all. In this picture the **g** is the lady in purple who is holding the yellow sign. Children will rely on visual imprinting to remember the spelling of this word.

Write SIGN on your whiteboard and add two compound words: "align, resign, design, assign, benign, cosign, malign." There are other words that use the GN spelling with a long A vowel sound: "reign, feign, arraign, foreign, campaign, Champaign, sovereign."

Children will make up a sentence they like, and then write and illustrate it. Ex: "I need to design a sign that will align well." "The sign did not align, so I will resign."

List N1, Level 5

stake

Show the children the SnapWord® card for STAKE. Study the picture together and ask the children to comment on what they see. Stake sounds just like the steak we eat, so we will need to focus on the elements in the picture that help us remember the spelling and meaning of this word. The **s** is a girl who is pounding in a tent stake. The **t** is right at the opening to the tent. Notice the **k** which is a pine tree. What is left is an **a** and the final pinchy **e** who is making **a** say her name.

Follow the routine next: Body motion, sentence, visualization, and writing.

Write STAKE on your whiteboard and add two compound words: "make, take, cake, lake, bake, quake, rake, wake, brake, shake, snake."

Children will make up a sentence they like, and then write and illustrate it. Ex: "I will take a stake to the lake." "I see a snake by the stake at the lake."

eyes

Show the children the SnapWord® card for EYES. Study the picture together and ask the children to comment on what they see. This picture will cement visually for the child how to spell EYES. Each **e** is an eye and the **y** suggests a nose. The final **s** signifies we are speaking of two eyes, not one eye. This spelling is unique for the long **i** sound. Normally **ey** says long **a** or long **e** (prey and monkey).

Follow the routine next: Body motion, sentence, visualization, and writing.

Write EYES on your whiteboard and add: "oxeye, eyelid, eyeball, eyeful, redeye, eyeing, bugeye, pinkeye, sockeye, walleye, buckeye, fisheye, eyesore."

Children will make up a sentence they like, and then write and illustrate it. Ex: "The fish's eyes are an eyesore!" "Some fish have EYES in their names – walleyes and sockeyes."

hair

Show the children the SnapWord® card for HAIR. Study the picture together and ask the children to comment on what they see. This picture clearly shows the meaning of the word. The girl who is the **i** is even looking surprised at the cloud of hair she has. The mirror looks much too small for her to use as she combs her hair. Notice that her elbow is resting on the **a** and her pink collar is touching the **r**.

Follow the routine next: Body motion, sentence, visualization, and writing.

Write HAIR on your whiteboard and add two compound words: "lair, fair, lair, pair, chair, flair, stair, air."

Children will make up a sentence they like, and then write and illustrate it. Ex: "Her hair is in the air with flair!" "The pair on the chair have big hair."

List N1, Level 5

chair

Show the children the SnapWord® card for CHAIR. Study the picture together and ask the children to comment on what they see. This picture clearly shows the meaning of the word and it helps with remembering the spelling. The **h** is the red chair while the **a** is the footstool. The **i** is the girl who is excited to show off her new chair.

Follow the routine next: Body motion, sentence, visualization, and writing.

Write CHAIR on your whiteboard and refer to the word list from HAIR.

Children will make up a sentence they like, and then write and illustrate it. Ex: "The chair at the fair is by the stair." "I like to sit on my chair when the air is fair."

world

Show the children the SnapWord® card for WORLD. Study the picture together and ask the children to comment on what they see. This picture clearly shows the meaning of the word and also provides a visual prompt so the child will remember that **er** in this word is spelled **or**. Talk about the placement of the planet and the stars around the world. The **l** is tipping towards the planet as though wanting to see it better. The letters in WORLD are tippy and look like they are floating in space.

Follow the routine next: Body motion, sentence, visualization, and writing.

Write WORLD on your whiteboard and add: "worth, words, worry, worst, visitor, color, elevator, director, navigator." Break the longer words into syllables. Each syllable has simple letter sounds, so if children are listening for them, they will be able to write what they are hearing/saying. Ex: VI-SI-TOR. E-LE-VA-TOR. DI-REC-TOR. NA-VI-GA-TOR.

Children will make up a sentence they like, and then write and illustrate it. Ex: "On the map of the world, the colors are good." "The worst visitor in the world came today!" "I will not worry about the world."

clothes

Show the children the SnapWord® card for CLOTHES. Study the picture together and ask the children to comment on what they see. OH MY! Look at all the clothes that need to be put away! Pay special attention to the red shoes in the **e** because this is the letter many people omit when they write clothes. If you leave out the **e**, you have cloths – like rags you dust with. Ask the children to talk about what they see in each letter. Rely on visual imprinting with this one – with their eyes closed, can they "see" the clothes on the various letters?

Follow the routine next: Body motion, sentence, visualization, and writing.

Write CLOTHES on your whiteboard and add: "bathes, seethes, loathes, breathes, writhes, soothes, tithes."

Children will make up a sentence they like, and then write and illustrate it. Ex: "When he bathes his dog, he loathes it when his clothes get wet." "It soothes her when her clothes are not too small."

List N1, Level 5

snow

Show the children the SnapWord® card for SNOW. Study the picture together and ask the children to comment on what they see. This picture clearly shows the meaning of the word! In fact some of the edges of the letters are snow covered. Note the evergreen tree between the **s** and the **n**. The redbird perches on the **o** and the **w** is along for the ride, combining with the **o** to make the OH sound.

Follow the routine next: Body motion, sentence, visualization, and writing.

Write SNOW on your whiteboard and add two compound words: "show, low, blow, know, crow, flow, glow, grow, borrow, below, bestow, throw, outgrow, willow."

Children will make up a sentence they like, and then write and illustrate it. Ex: "The mole will burrow down low below the snow." "I know the willow is below the snow."

children

Show the children the SnapWord® card for CHILDREN. Study the picture together and ask the children to comment on what they see. Approach this word in its two syllables. The first syllable is actually a whole word: CHILD. The two older children are in CHILD, while the small children are in **ren**. The placement of the two colored balls highlight the vowels in each syllable: **i** and **e**.

Follow the routine next: Body motion, sentence, visualization, and writing.

Write CHILDREN on your whiteboard and add: "wren, silken, hidden, linden, drench, stepchildren, godchildren, grandchildren."

Children will make up a sentence they like, and then write and illustrate it. Ex: "The children saw the wren hidden in the linden." "These children have silken hair." "These are my stepchildren and grandchildren."

year

Show the children the SnapWord® card for YEAR. Study the picture together and ask the children to comment on what they see. This picture shows the four seasons in the year: summer, spring, fall and winter. Let the children talk about the way the four seasons fit with each letter. The **y** is a great tree trunk, for example. The **e** is round like the sun, the **a** fits right on the fall leaf and the **r** is snow-covered.

Follow the routine next: Body motion, sentence, visualization, and writing.

Write YEAR on your whiteboard and add: "fear, ear, tear (like crying), smear, shear, near, clear, spear."

Children will make up a sentence they like, and then write and illustrate it. Ex: "I fear my ear is near the spear this year." "The new year is near – time to shear the sheep!"

List N1, Level 5

street

Show the children the SnapWord® card for STREET. Study the picture together and ask the children to comment on what they see. In this picture, the word STREET is lying flat on the street. The two **e**'s are working together to say long **e**. Find the word TREE inside STREET and point out that all you do is add an **s** at the beginning and a **t** at the end.
    Follow the routine next: Body motion, sentence, visualization, and writing.
    Write STREET on your whiteboard and add: "tweet, sweet, sheet, greet, feet, fleet, meet."
    Children will make up a sentence they like, and then write and illustrate it. Ex: "I will tweet about the sheet on the street." "We will meet and greet the fleet coming up the street."

rock

Show the children the SnapWord® card for ROCK. Study the picture together and ask the children to comment on what they see. In this picture, the **o** is the rock and the **k** is the girl who just spotted a frog sitting on the rock. The **ck** work together to make the K sound.
    Follow the routine next: Body motion, sentence, visualization, and writing.
    Write ROCK on your whiteboard and add two compound words: "stock, crock, lock, frock, smock, clock, shock, block."
    Children will make up a sentence they like, and then write and illustrate it. Ex: "I put my block in a crock on the rock." "I will lock the crock by the rock."

sand

Show the children the SnapWord® card for SAND. Study the picture together and ask the children to comment on what they see. The letters in this picture look like they are made of sand. They are sagging a bit. Find the small words inside of sand together: A, AN, AND.
    Follow the routine next: Body motion, sentence, visualization, and writing.
    Write SAND on your whiteboard and add two compound words: "stand, strand, band, bland, hand, land, command."
    Children will make up a sentence they like, and then write and illustrate it. Ex: "I like to stand on my hand in the sand." "I will command the band to play on the strand of sand."

SNAPWORDS LIST N2

List N2 Words - ABC order:	List N2 Words by level:
air baby bike boat body boot bus class dad dinner dream earth fact family fare feet field fish game group heart hill hour idea job kids lady land life line list lunch Mom moon nothing number park party past person picture place problem reason sea ship sky space spring state story stuff summer things time trouble week winter yard	**Level 1:** bus, dad, fact, fish, hill, job, kids, land, list, mom, past, ship **Level 2:** baby, bike, body, class, game, idea, lady, life, line, lunch, place, sky **Level 3:** boat, boot, dinner, dream, family, feet, moon, sea, space, state, stuff, time **Level 4:** air, field, hour, park, party, spring, story, summer, things, week, winter, yard **Level 5:** earth, fare, group, heart, nothing, number, person, picture, problem, reason, trouble,

How To Teach   List N2, Level 1

bus

Show the children the SnapWord® card for BUS. Study the picture together and ask the children to comment on what they see. The back of the **b** becomes the windshield and frames the face of the first child on the bus. This word is very easy to sound out - the primary challenge for some children might be to not reverse the **b**. Studying the picture will help with this.

Follow the routine next: Do the body motion from the back of the card, read the sentence, close eyes and do visualization (see the word in ones mind), and practice writing the word on whiteboards.

Write BUS on your whiteboard. Add: "bus, plus, pus, thus."   ** In this section of the Lessons, always write the words in a column so children will be able to see the sound spelling the words have in common.

Children will make up a sentence they like, and then write and illustrate it. Ex: "I see a bus plus some kids."

dad

Show the children the SnapWord® card for DAD. Study the picture together and ask the children to comment on what they see. This word is also very easy to read and write. Notice the two **d**'s are the dads and they are "twins." The **a** is just like them except for the missing tall part of the letter **d**.

Follow the routine next: Body motion, sentence, visualization, and writing.

Write DAD on your whiteboard and add: "bad, fad, had, lad, mad, pad, sad, tad, clad, glad."

Children will make up a sentence they like, and then write and illustrate it. Ex: "Dad was glad to have a pad!"

fact

Show the children the SnapWord® card for FACT. Study the picture together and ask the children to comment on what they see. The teacher (letter **t**) is a teacher who is writing a fact on the blackboard while the student (letter **a**) watches. This word can be sounded out, but studying the picture will bypass the need to sound it out.

Follow the routine next: Body motion, sentence, visualization, and writing.

Write FACT on your whiteboard and add: "act, pact, tact, tract."

Sample sentences: "We made a pact to act with tact."

List N2, Level 1

fish

   Show the children the SnapWord® card for FISH. Study the picture together and ask the children to comment on what they see. Our girl just caught a fish and she looks as surprised as the fish! The only part of this word that isn't straightforward is the digraph **sh**.
   Follow the routine next: Body motion, sentence, visualization, and writing.
   Write FISH on your whiteboard and add: "dish, wish, swish."
   Children will make up a sentence they like, and then write and illustrate it. Ex: "I wish I had a fish in my dish."

hill

   Show the children the SnapWord® card for HILL. Study the picture together and ask the children to comment on what they see. The letters are lying down on a hill. The **i** is right on the road while the tall letters are on either side of it. The only spelling trick is the double **l**.
   Follow the routine next: Body motion, sentence, visualization, and writing.
   Write HILL on your whiteboard and add: "bill, dill, fill, gill, Jill, kill, mill, pill, quill, rill, sill, till, will, chill, drill, frill, grill, skill, spill, still, swill, thrill, trill, twill."
   Children will make up a sentence they like, and then write and illustrate it. Ex: "Bill has a mill on the hill."

job

   Show the children the SnapWord® card for JOB. Study the picture together and ask the children to comment on what they see. This picture is very interesting! The **j** is a lady who is cleaning the toilet (the **o**) and the **b** at the end is holding all her cleaning supplies.
   Follow the routine next: Body motion, sentence, visualization, and writing.
   Write JOB on your whiteboard and add: "cob, fob, gob, knob, lob, mob, rob, sob, blob, glob, slob, snob."
   Children will make up a sentence they like, and then write and illustrate it. Ex: "I will not sob over my job!"

List N2, Level 1

kids

Show the children the SnapWord® card for KIDS. Study the picture together and ask the children to comment on what they see. This picture shows two kinds of kids: a child and a baby goat. They are the short letters. The tall letters, **k** and **d** are the parents.

Follow the routine next: Body motion, sentence, visualization, and writing.

Write KIDS on your whiteboard and add: "bids, rids, acids, grids, skids."

Children will make up a sentence they like, and then write and illustrate it. Ex: "The kids made grids from the skids."

land

Show the children the SnapWord® card for LAND. Study the picture together and ask the children to comment on what they see. This picture shows two meanings of the word LAND. A plane is about to land on a piece of land. Look for smaller words "AN" and "AND" inside LAND.

Follow the routine next: Body motion, sentence, visualization, and writing.

Write LAND on your whiteboard. Also write: "and, band, hand, sand, bland, brand, gland, stand, strand, grand."

Children will make up a sentence they like, and then write and illustrate it. Ex: "The band will stand on the land to play."

list

Show the children the SnapWord® card for LIST. Study the picture together and ask the children to comment on what they see. There is a huge, long list of chores and it looks like the boy is surprised and shocked. Maybe he's deciding which to start with.

Follow the routine next: Body motion, sentence, visualization, and writing.

Write LIST on your whiteboard and add: "fist, mist, wrist, grist, twist, exist, insist, typist, artist, resist, assist, enlist."

Children will make up a sentence they like, and then write and illustrate it. Ex: "I will twist the list in my fist."

List N2, Level 1

mom

Show the children the SnapWord® card for MOM. Study the picture together and ask the children to comment on what they see. In this picture, Mom is the capital **M** and her baby is the **o**. Mom is wiping a runny nose.

Follow the routine next: Body motion, sentence, visualization, and writing.

Write MOM on your whiteboard and add: "from, atom, bottom, custom, seldom, sitcom, fathom, random, wisdom."

Children will make up a sentence they like, and then write and illustrate it. Ex: "Mom likes a random sitcom." Or, "There is a random atom on the bottom."

past

Show the children the SnapWord® card for PAST. Study the picture together and ask the children to comment on what they see. This picture shows two meanings of the word PAST. One meaning is something that happened in the PAST or long ago (like using covered wagons in the past). The other meaning is that something is going by. The cars are going PAST the word. This word can be sounded out. The **st** ending is the same as in the word LIST.

Follow the routine next: Body motion, sentence, visualization, and writing.

Write PAST on your whiteboard and add: "cast, last, vast, fast, mast, outlast, webcast."

Children will make up a sentence they like, and then write and illustrate it. Ex: "In the past, the cast was vast."

ship

Show the children the SnapWord® card for SHIP. Study the picture together and ask the children to comment on what they see. This picture shows a ship floating in the sea, anchor dropped to hold it in place. The **h** and the **i** are laid over towers on the ship. The digraph **sh** is the only sound spelling to pay attention to.

Follow the routine next: Body motion, sentence, visualization, and writing.

Write SHIP on your whiteboard and add: "dip, hip, lip, nip, quip, rip, sip, tip, zip, blip, chip, clip, drip, flip, grip, skip, slip, snip, strip, trip, whip."

Children will make up a sentence they like, and then write and illustrate it. Ex: "He will skip the trip on the ship."

List N2, Level 2

baby

Show the children the SnapWord® card for BABY. Study the picture together and ask the children to comment on what they see. The little, round **a** is the baby and the second tall letter is Grandma. Notice that the flat part of the **b**'s and the **a** are on different sides of the letters. The flat sides of the baby and the grandma are facing each other. This will matter for children who tend to reverse lowercase **b**. The **y** at the end sounds like long **e**.

Follow the routine next: Body motion, sentence, visualization, and writing.

Write BABY on your whiteboard and add: "rugby, derby." Next, add these words that have a double B: "nubby, hobby, webby, chubby, knobby, shabby, stubby." ** In this section of the Lessons, always write the words in a column so children will be able to see the sound spelling the words have in common.

Children will make up a sentence they like, and then write and illustrate it. Ex: "The baby is chubby."

bike

Show the children the SnapWord® card for BIKE. Study the picture together and ask the children to comment on what they see. Notice the letters that make the bike wheels - the **b** and the final **e**. It is convenient that the back of the **b** is there to put the handle bars on. In this word, the sound spelling is long **i** with the Pinchy **e**.

Follow the routine next: Body motion, sentence, visualization, and writing.

Write BIKE on your whiteboard and add: "like, Mike, dike, pike, spike, strike." There are many other words with the long I, pinchy E spelling but with a different consonant.

Children will make up a sentence they like, and then write and illustrate it. Ex: "Mike likes to ride his bike on the dike."

body

Show the children the SnapWord® card for BODY. Study the picture together and ask the children to comment on what they see. I love this picture because the bears are so blobby and lumpy. Also, best of all, they are facing each other which will help children write the letters correctly. The bears' bellies are facing each other. The only sound on which to focus is the final **y** that sounds like a long **e** - like in BABY.

Follow the routine next: Body motion, sentence, visualization, and writing.

Write BODY on your whiteboard and add these great compound body words: "embody, nobody, anybody, busybody, somebody, underbody, everybody."

Children will make up a sentence they like, and then write and illustrate it. Ex: "Does everybody want to help me paint the underbody of my car? Nobody? Anybody?"

List N2, Level 2

class

Show the children the SnapWord® card for CLASS. Study the picture together and ask the children to comment on what they see. Oh my goodness! This class is learning how to knit and they are all making scarves of different colors. The tall letter **l**, is the teacher. The only sound spelling on which to focus is the final **s** which is double.

Follow the routine next: Body motion, sentence, visualization, and writing.

Write CLASS on your whiteboard and add: "glass, lass, brass, crass, mass, pass, bass."

Children will make up a sentence they like, and then write and illustrate it. Ex: "The lass took a class after mass."

game

Show the children the SnapWord® card for GAME. Study the picture together and ask the children to comment on what they see. In this picture, friends (the **g** and the **e**) are playing a game on a low table that is made of the little word AM in the middle of GAME. The sound spelling is long **a** with a Pinchy **e**.

Follow the routine next: Body motion, sentence, visualization, and writing.

Write GAME on your whiteboard and add: "came, dame, fame, lame, name, same, tame, blame, flame, frame, shame."

Children will make up a sentence they like, and then write and illustrate it. Ex: "The same dame came to play a game."

idea

Show the children the SnapWord® card for IDEA. Study the picture together and ask the children to comment on what they see. This picture shows clearly what it looks like when someone gets an idea. The word IDEA is simple. The **i** and the **e** are both long, and then the final **a** sounds like "uh."

Follow the routine next: Body motion, sentence, visualization, and writing.

Write IDEA on your whiteboard and add: "urea, spirea, chorea, cornea, apnea, cornea, nausea, azalea."

Children will make up a sentence they like, and then write and illustrate it. Ex: "She got the idea to plant spirea and azalea bushes."

List N2, Level 2

lady

Show the children the SnapWord® card for LADY. Study the picture together and ask the children to comment on what they see. This picture shows a lady who seems to enjoy having scarves. The **l** is the scarf holder, the **d** is the lady herself, and the **y** is holding a green scarf she might be planning to wear. This word is similar to BABY, with its long **a** and final **y** that sounds like long **e**. The sound of final **y** is the same as in BODY also.

Follow the routine next: Body motion, sentence, visualization, and writing.

Write LADY on your whiteboard and add: "shady, handy, dandy, candy, howdy, rowdy, tardy, hardy, wordy."

Children will make up a sentence they like, and then write and illustrate it. Ex: "The lady gave the rowdy kids some candy."

life

Show the children the SnapWord® card for LIFE. Study the picture together and ask the children to comment on what they see. I love this picture! It looks so full of life, doesn't it? There are many living things. Ask the children to identify them. It looks like the boy in the picture loves it! The sound spelling for LIFE is long **i** with a Pinchy **e**.

Follow the routine next: Body motion, sentence, visualization, and writing.

Write LIFE on your whiteboard and add: "fife, knife, rife, wife, strife."

Children will make up a sentence they like, and then write and illustrate it. Ex: "His wife said there is no strife in life!"

line

Show the children the SnapWord® card for LINE. Study the picture together and ask the children to comment on what they see. This picture shows kids in line going to recess. Each child is carrying a ball to play with as they follow the teacher. The focus sound spelling in this word is the long **i** with Pinchy **e** spelling as we saw in LIFE and BIKE.

Follow the routine next: Body motion, sentence, visualization, and writing.

Write LINE on your whiteboard and add: "dine, fine, mine, nine, pine, tine, vine, wine, brine, shine, shrine, spine, swine, whine."

Children will make up a sentence they like, and then write and illustrate it. Ex: "Nine of us got in line to go dine." OR "Nine of us in line did not whine!"

List N2, Level 2

lunch

Show the children the SnapWord® card for LUNCH. Study the picture together and ask the children to comment on what they see. This picture shows two people having a picnic. Notice what each of the letters are. The man appears to have an enormous drink, while the lady has a large orange. Maybe they plan to share the sandwich that is behind the **n**. There is no tricky spelling to this word. Just the final digraph, **ch**.

Follow the routine next: Body motion, sentence, visualization, and writing.

Write LUNCH on your whiteboard and add: "bunch, hunch, munch, punch, brunch, crunch."

Children will make up a sentence they like, and then write and illustrate it. Ex: "I have a hunch we will munch and crunch at lunch!"

place

Show the children the SnapWord® card for PLACE. Study the picture together and ask the children to comment on what they see. In this picture, there is a line of kids behind their teacher, and then one boy (the **p**) who is in his place to one side of the rest. The spellings in this word to notice are long **a** with Pinchy **e** and the **ce** /S/ spelling.

Follow the routine next: Body motion, sentence, visualization, and writing.

Write PLACE on your whiteboard and add: "face, lace, mace, pace, race, brace, grace, space, trace."

Children will make up a sentence they like, and then write and illustrate it. Ex: "She had the grace to race in place." OR "I see a trace of lace about her face."

sky

Show the children the SnapWord® card for SKY. Study the picture together and ask the children to comment on what they see. This picture shows a busy sky! There is a moon, a jet, a waving flag and several birds - oh and two clouds - occupying the space in the sky. In this easy word, the spelling is the final **y** that sounds like long **i**. NOTE: when words end in **y** that are more than one syllable, the **y** will sound like long **e**.

Follow the routine next: Body motion, sentence, visualization, and writing.

Write SKY on your whiteboard and add: "pry, cry, sly, spy, fry, fly, spry, try, dry, wry." All these words are just one syllable, so the Y sounds like long I.

Children will make up a sentence they like, and then write and illustrate it. Ex: "I try to spy it fly in the sky."

List N2, Level 3

boat

Show the children the SnapWord® card for BOAT. Study the picture together and ask the children to comment on what they see. This picture shows a boat floating on the water. The three round letters each have a window inside them. The tall letters form smoke stacks. The sound spelling here is long **o** spelled **oa**.
Follow the routine next: Body motion, sentence, visualization, and writing.
Write BOAT on your whiteboard and add: "coat, goat, moat, oat, bloat, gloat, float, throat." ** In this section of the Lessons, always write the words in a column so children will be able to see the sound spelling the words have in common.
Children will make up a sentence they like, and then write and illustrate it. Ex: "The goat in the coat will float in the moat."

boot

Show the children the SnapWord® card for BOOT. Study the picture together and ask the children to comment on what they see. This is one giant boot! the **b** at the beginning fits in the tall part of the boot, while the **oo** is the toe. The final **t** is the cowboy who looks very surprised at the size of the boot. The focus sound spelling is **oo** that sounds like the **oo** in BOO.
Follow the routine next: Body motion, sentence, visualization, and writing.
Write BOOT on your whiteboard and add: "hoot, loot, moot, root, toot, scoot, shoot, booth, tooth, broom, groom, room."
Children will make up a sentence they like, and then write and illustrate it. Ex: "It was a hoot to scoot by the booth."

dinner

Show the children the SnapWord® card for DINNER. Study the picture together and ask the children to comment on what they see. I like this picture because of the way the letters easily form Mom and child and then the table. It looks like Mom is bringing food to the table and her child is getting ready to set another blue glass on the table. The only sound spellings to notice in DINNER are the double **n** in the middle, and Bossy **r** (ER) ending.
Follow the routine next: Body motion, sentence, visualization, and writing.
Write DINNER on your whiteboard and add: "winner, thinner, fanner, banner, manner, funner, runner."
Children will make up a sentence they like, and then write and illustrate it. Ex: "After dinner the runner was a winner."

List N2, Level 3

dream

Show the children the SnapWord® card for DREAM. Study the picture together and ask the children to comment on what they see. In the picture we see a girl just waking up from a dream about a bear. I can imagine she is so happy to be awake! The sound spelling in this word is long **e** spelled **ea** like it is in EAT.

Follow the routine next: Body motion, sentence, visualization, and writing.

Write DREAM on your whiteboard and add: "beam, ream, seam, cream, gleam, scream, steam, stream, team."

Children will make up a sentence they like, and then write and illustrate it. Ex: "In my dream, the team was in the stream."

family

Show the children the SnapWord® card for FAMILY. Study the picture together and ask the children to comment on what they see. This picture looks so busy! It is playtime in the family and they are playing with two balls and a pull toy duck. Mom and Dad are the tall letters. The final **y** sounds like long **e** (because the word is multi-syllable.)

Follow the routine next: Body motion, sentence, visualization, and writing.

Write FAMILY on your whiteboard and add: "daily, gaily, tidily, lazily, cosily, bodily, nosily, busily, easily." All of these words started as simpler words ending in **y** but were changed by adding **ily** ending to tell how something was being done. Example: "noisy" changed to "noisily" which tells how they were doing it. They were doing it noisily. "Easy" changes to "easily" which also tells how they were doing it. They were doing it easily. In each case, remove the **y** and add **ily**. "Day" turns into "daily" which tells how often they do something. They do it daily.

Children will make up a sentence they like, and then write and illustrate it. Ex: "Daily, the family played noisily."

feet

Show the children the SnapWord® card for FEET. Study the picture together and ask the children to comment on what they see. This picture shows four pairs of bare feet. The family is obviously checking out the different sizes of feet in their family. The tall letters are the parents while the **e**'s are the kids. The sound spelling focus is long **e** spelled **ee**.

Follow the routine next: Body motion, sentence, visualization, and writing.

Write FEET on your whiteboard and add: "beet, meet, fleet, greet, sheet, skeet, sleet, street, sweet, tweet."

Children will make up a sentence they like, and then write and illustrate it. Ex: "When there is sleet in the street, we have shoes on our feet."

List N2, Level 3

moon

Show the children the SnapWord® card for MOON. Study the picture together and ask the children to comment on what they see. This picture with its two **oo**'s show the earth and the moon together in space. The target sound spelling is the long sound of **oo**.

Follow the routine next: Body motion, sentence, visualization, and writing.

Write MOON on your whiteboard and add: "loon, noon, soon, croon, spoon, swoon."

Children will make up a sentence they like, and then write and illustrate it. Ex: "Soon the loon will swoon in the light of the moon."

sea

Show the children the SnapWord® card for SEA. Study the picture together and ask the children to comment on what they see. In this picture, the smaller bodies of water found in between the letter parts are seas. There are little boats floating in the sea. The sound spelling is the same as in DREAM: Long **e** spelled **ea**.

Follow the routine next: Body motion, sentence, visualization, and writing.

Write SEA on your whiteboard and add: "tea, pea, flea, plea, lea." Of course many, many other words have the EA long E spelling, but these words end with EA.

Children will make up a sentence they like, and then write and illustrate it. Ex: "The flea had a pea and tea on a boat in the sea."

space

Show the children the SnapWord® card for SPACE. Study the picture together and ask the children to comment on what they see. This picture is similar to the picture for MOON. This time, the earth is in the front and the moon is tucked into the **c**. The sound spelling focus for SPACE is the **ce** ending that sounds like **s**. We studied this sound spelling in the lesson for PLACE.

Follow the routine next: Body motion, sentence, visualization, and writing.

Write SPACE on your whiteboard and add: "face, lace, mace, pace, race, brace, grace, place, trace."

Children will make up a sentence they like, and then write and illustrate it. Ex: "I set the pace in the race to outer space."

List N2, Level 3

state

Show the children the SnapWord® card for STATE. Study the picture together and ask the children to comment on what they see. This picture shows a map of the STATE of Michigan. The **t**'s are both kids wearing mittens - after all they are in the mitten state! The sound spelling focus is long **a** spelled **a-e**: **a** with a Pinchy **e**.

Follow the routine next: Body motion, sentence, visualization, and writing.

Write STATE on your whiteboard and add: "ate, date, fate, gate, hate, late, mate, rate, crate, grate, plate, skate."

Children will make up a sentence they like, and then write and illustrate it. Ex: "In my state we hate to be late to skate!"

stuff

Show the children the SnapWord® card for STUFF. Study the picture together and ask the children to comment on what they see. In this picture we see three boys gathering up all their STUFF to put into the little wagon they have. The only sound spelling is the double **ff**. In this picture the **ff**'s are dressed alike.

Follow the routine next: Body motion, sentence, visualization, and writing.

Write STUFF on your whiteboard and add: "cuff, huff, puff, ruff, bluff, fluff, gruff, scuff, sluff, snuff."

Children will make up a sentence they like, and then write and illustrate it. Ex: "We huff and puff as we put our stuff away."

time

Show the children the SnapWord® card for TIME. Study the picture together and ask the children to comment on what they see. In this picture, the boy is talking about the passing of time from day to night. Time passes as we go from sun to moon. The sound spelling focus is of course long **i** spelled **i-e** (Pinchy **e**).

Follow the routine next: Body motion, sentence, visualization, and writing.

Write TIME on your whiteboard and add: "dime, lime, mine, chime, clime, crime, grime, prime, slime."

Children will make up a sentence they like, and then write and illustrate it. Ex: "In time my prime lime turned to slime!"

List N2, Level 4

Show the children the SnapWord® card for AIR. Study the picture together and ask the children to comment on what they see. In this picture, the girl looks like she is taking a deep breath of the cool mountain AIR. The dot on the **i** is her open mouth. The sound spelling we will focus on is long **a** spelled **ai**.

Follow the routine next: Body motion, sentence, visualization, and writing.

Write AIR on your whiteboard and add: "fair, hair, lair, pair, chair, flair, stair." ** In this section of the Lessons, always write the words in a column so children will be able to see the sound spelling the words have in common.

Children will make up a sentence they like, and then write and illustrate it. Ex: "The pair went to the fair with their hair in the air."

field

Show the children the SnapWord® card for FIELD. Study the picture together and ask the children to comment on what they see. The picture for FIELD shows the **e** sound/letter being emphasized by being in the corner of the fence. The tricky spelling in this word is the long **e** which is spelled **ie**.

Follow the routine next: Body motion, sentence, visualization, and writing.

Write FIELD on your whiteboard and add: "brief, chief, grief, relief, niece, piece, siege, priest, fierce, believe, achieve, relieve, retrieve."

Children will make up a sentence they like, and then write and illustrate it. Ex: "I saw the chief, my niece, and a priest in the field!" Or "The chief and my niece are in the field."

hour

Show the children the SnapWord® card for HOUR. Study the picture together and ask the children to comment on what they see. The picture for HOUR shows the silent **h** as a man checking his watch - separate from the rest of the word (OUR). The **o** is the lid of the soup pot which is the **u**. The **R** is the table holding a bowl with a spoon. The two spellings on which to focus are the silent **h** and the **ou** spelling that sounds like **ow**.

Follow the routine next: Body motion, sentence, visualization, and writing.

Write HOUR on your whiteboard and add "our, sour, flour, scour." Other words with silent H include: "honest, honor, ghost, ghastly, gherkin, ghetto, rhino, rhubarb, rhythm, exhibit."

Children will make up a sentence they like, and then write and illustrate it. Ex: "In one hour, our flour will be sour!" And "In one hour, the exhibit of rhubarb and gherkins will begin." "The ghost was in the ghetto for an hour."

List N2, Level 4

park

Show the children the SnapWord® card for PARK. Study the picture together and ask the children to comment on what they see. This picture shows two meanings of park: to park a car in the park. The focus sound spelling is **ar** that sounds like the name of the letter **r**. The next lesson (for PARTY) also follows this sound spelling pattern and you could introduce the words together if desired.

Follow the routine next: Body motion, sentence, visualization, and writing.

Write PARK on your whiteboard and add: "ark, bark, hark, lark, mark, shark, spark, stark."

Children will make up a sentence they like, and then write and illustrate it. Ex: "My dog will bark at a lark in the park."

party

Show the children the SnapWord® card for PARTY. Study the picture together and ask the children to comment on what they see. This picture shows a party with two people chatting, taking a drink, and apparently planning to eat some pink cupcakes! The sound spellings include **ar** as also in PARK, and the final **y** that sounds like Long **e**.

Follow the routine next: Body motion, sentence, visualization, and writing.

Write PARTY on your whiteboard and use the word list from PARK. Also check word lists from Section 2 for BABY, BODY and LADY.

Children will make up a sentence they like, and then write and illustrate it. Ex: "At the party in the city she was chatty." "Hark, I hear a lark in the party in the park!"

spring

Show the children the SnapWord® card for SPRING. Study the picture together and ask the children to comment on what they see. The picture for SPRING looks like spring for sure! Talk about what is on or in each letter of the word. The final **ing** spelling is the focus here because all the rest of the sounds can be easily heard and therefore written.

Follow the routine next: Body motion, sentence, visualization, and writing.

Write SPRING on your whiteboard and add: "bling, sling, string, bring, ring, king, sting, ping, sing, wing, zing, cling, fling, swing, thing, wring."

Children will make up a sentence they like, and then write and illustrate it. Ex: "In the spring, I swing, bees sting, and birds sing." "The king will bring a sling and string."

List N2, Level 4

story

Show the children the SnapWord® card for STORY. Study the picture together and ask the children to comment on what they see. This picture makes me feel as though I am there listening to Dad tell his story. The sound spelling on which to focus is **or**. We have already talked about final **y** several times.

Follow the routine next: Body motion, sentence, visualization, and writing.

Write STORY on your whiteboard and add: "for, cord, ford, scorch, sword, cork, stork, dorm, storm, scorn, sworn, thorn, short, snort, sport."

Children will make up a sentence they like, and then write and illustrate it. Ex: "His story was about a short stork in a storm."

summer

Show the children the SnapWord® card for SUMMER. Study the picture together and ask the children to comment on what they see. The picture helps us focus on the two sound spellings of note. One is the double **m** (there is a towel hanging from the second **m**) and the final **er** (which is sheltered from the sun by the pink umbrella).

There is a shovel in the first vowel - **u** - just in case there is any confusion about which letter makes that sound.

Follow the routine next: Body motion, sentence, visualization, and writing.

Write SUMMER on your whiteboard and add: "bummer, hummer." There are also many words with double M and final ER: yammer, rammer, simmer, jammer, dimmer, shimmer, stammer, slimmer, swimmer, grimmer, hammer, trimmer."

Children will make up a sentence they like, and then write and illustrate it. Ex: "The swimmer is slimmer in the summer." "In summer, the sun will shimmer on the swimmer."

things

Show the children the SnapWord® card for THINGS. Study the picture together and ask the children to comment on what they see. In this picture we can see all the things the boys need in order to play a game of baseball. The two sound spellings in this word are pretty familiar: the digraph **th** and the ending **ing**.

Follow the routine next: Body motion, sentence, visualization, and writing.

Write THINGS on your whiteboard and refer to the word list from the lesson on SPRING.

Children will make up a sentence they like, and then write and illustrate it. Ex: "In the spring, she made a swing from string."

List N2, Level 4

week

Show the children the SnapWord® card for WEEK. Study the picture together and ask the children to comment on what they see. In this picture, a student is pointing to a week (7 days) on a calendar and the teacher is clapping for her! The sound spelling is of course long **e** spelled **ee**.

Follow the routine next: Body motion, sentence, visualization, and writing.

Write WEEK on your whiteboard and add: "leek, meek, peek, reek, seek, cheek, creek, Greek, sleek, bleed, breed, creed, freed, greed, speed, steed, tweed, sheep, wheel, street" and so many more!

Children will make up a sentence they like, and then write and illustrate it. Ex: "This week my sheep were freed with speed!" "This week I will peek at the sleek sheep in the creek!"

winter

Show the children the SnapWord® card for WINTER. Study the picture together and ask the children to comment on what they see. Not everyone enjoys snow in the winter, but these kids are! They are sliding down a hill on their long, red sled. Notice the small word IN and then the final **er** ending. Another small word in WINTER is WIN.

Follow the routine next: Body motion, sentence, visualization, and writing.

Write WINTER on your whiteboard and add: "printer, splinter, sprinter."

Children will make up a sentence they like, and then write and illustrate it. Ex: "The sprinter got a splinter last winter!"

yard

Show the children the SnapWord® card for YARD. Study the picture together and ask the children to comment on what they see. In this picture we see a happy back yard. The grass is a little long, there is a bird nest in the **y**, and the orange snake looks right at home! The focus is the sound spelling **ar**. Reference lessons for PARK and PARTY.

Follow the routine next: Body motion, sentence, visualization, and writing.

Write YARD on your whiteboard and add: "bard, card, chard, gard, hard, lard, shard, buzzard, regard."

Children will make up a sentence they like, and then write and illustrate it. Ex: "A bard got chard from the yard." "It is not hard to cook chard in lard in the back yard."

List N2, Level 5

earth

Show the children the SnapWord® card for EARTH. Study the picture together and ask the children to comment on what they see. In this picture we see the word earth superimposed on the globe. We can see water and land as well as the relationship of the earth to the sun and planets. The word EARTH has a smaller word in it: EAR. The sound spelling focus is on the sound of **er** spelled **ear**.

Follow the routine next: Body motion, sentence, visualization, and writing.

Write EARTH on your whiteboard and add: "heard, learn, earn, early, earnest, search, rehearse." ** In this section of the Lessons, always write the words in a column so children will be able to see the sound spelling the words have in common.

Children will make up a sentence they like, and then write and illustrate it. Ex: "I will learn what I heard about the earth."

fare

Show the children the SnapWord® card for FARE. Study the picture together and ask the children to comment on what they see. This picture shows a cowboy holding his FARE for a taxi ride. The smaller word inside of FARE is ARE, but it sounds like AIR. This is our sound spelling focus for this lesson.

Follow the routine next: Body motion, sentence, visualization, and writing.

Write FARE on your whiteboard and add: "hare, care, bare, rare, scare, share, stare, spare."

Children will make up a sentence they like, and then write and illustrate it. Ex: "The hare will care if he's bare so don't stare!"

group

Show the children the SnapWord® card for GROUP. Study the picture together and ask the children to comment on what they see. This picture clearly shows the meaning of the word. Each letter is a person in the discussion group. The focus sound spelling for GROUP is the long **oo** sound (as in MOON) but spelled **ou** as in YOU.

Follow the routine next: Body motion, sentence, visualization, and writing.

Write GROUP on your whiteboard and add two compound words: "you, soup, routine, youth, coupon, goulash, acoustic, wound, troupe."

Children will make up a sentence they like, and then write and illustrate it. Ex: "The youth group has a coupon for soup or goulash."

List N2, Level 5

heart

Show the children the SnapWord® card for HEART. Study the picture together and ask the children to comment on what they see. This picture shows what the word means, but also helps to break up the weird spelling a bit. The word HEART looks like a combination of two smaller words: HE and ART. The sound spelling is found in very few other words: **ar** spelled **ear**.

Follow the routine next: Body motion, sentence, visualization, and writing.

Write HEART on your whiteboard and add: "hearth, hearken."

Children will make up a sentence they like, and then write and illustrate it. Ex: "My heart will hearken as I sit on the hearth."

nothing

Show the children the SnapWord® card for NOTHING. Study the picture together and ask the children to comment on what they see. Oh my. This picture looks a bit sad, doesn't it? Apparently the kids were eating chips and ate them all. Now there is nothing left! The word NOTHING is made of two words NO and THING. Point this out and say, "There is NO THING left in the bowl!" Of course we say it like this: "NU-THING." The sound of the **o** as it says "**uh**" is our sound spelling focus for today.

Follow the routine next: Body motion, sentence, visualization, and writing.

Write NOTHING on your whiteboard and add: "done, mother, brother, other, some, come, shove, above, dove, love, coming, something."

Children will make up a sentence they like, and then write and illustrate it. Ex: "My mother and other brother are coming with something."

number

Show the children the SnapWord® card for NUMBER. Study the picture together and ask the children to comment on what they see. In this picture, the teacher is quizzing her students on which symbols are numbers. The smiling child picked up a number (the 5) while the other child chose a **t**, which is not a number. That's ok! He will choose a number next time! Our focus is simply on the final er ending. The word family we will look at is the BER family.

Follow the routine next: Body motion, sentence, visualization, and writing.

Write NUMBER on your whiteboard and add: "amber, ember, fiber, omber, saber, tuber, cyber, sober, umber, somber, timber, barber."

Children will make up a sentence they like, and then write and illustrate it. Ex: "A number of sober barbers have sabers."

List N2, Level 5

person

Show the children the SnapWord® card for PERSON. Study the picture together and ask the children to comment on what they see. In this picture, there is one PERSON and a few animals. The boy is saying he is the only PERSON present. The sound spelling focus is on **er** (which is a review from the lesson on NUMBER), while the word family is SON.

Follow the routine next: Body motion, sentence, visualization, and writing.

Write PERSON on your whiteboard and add: "mason, arson, bison, boson, reason, lesson, season, poison, prison, parson, unison, treason, venison."

Children will make up a sentence they like, and then write and illustrate it. Ex: "The person with bison and venison is Mason."

picture

Show the children the SnapWord® card for PICTURE. Study the picture together and ask the children to comment on what they see. In this picture we see a very shy person who doesn't want his picture taken! There is also a photographer, and one more person who DOES want his picture taken! The focus sound spelling in this lesson is the **ure** that sounds like **er** and the word family TURE.

Follow the routine next: Body motion, sentence, visualization, and writing.

Write PICTURE on your whiteboard and add: "capture, nature, creature, culture, feature, fracture, lecture, mixture, moisture, posture, puncture, structure, pasture, texture, picture, vulture."

Children will make up a sentence they like, and then write and illustrate it. Ex: "The picture will capture the creature in the pasture."

problem

Show the children the SnapWord® card for PROBLEM. Study the picture together and ask the children to comment on what they see. This picture shows four kids trying to solve a big problem. The l in the word is wobbling but it is stuck in a hole. Two of the kids are working to get it unstuck. Every part of this word could be sounded out, so there is nothing tricky. Each letter makes its own distinct sound. The word family today is EM.

Follow the routine next: Body motion, sentence, visualization, and writing.

Write PROBLEM on your whiteboard and add: "emblem, gem, hem, rem, ahem, item, poem, them, modem, harem, totem, anthem, system, stem."

Children will make up a sentence they like, and then write and illustrate it. Ex: "The problem with the item was the modem."

List N2, Level 5

reason

Show the children the SnapWord® card for REASON. Study the picture together and ask the children to comment on what they see. This picture shows a girl with her two dogs, one of which is marking the **a** that we don't hear in the word, the other which is digging a hole to bury its bone. The picture shows clearly the **re** that we hear when we say REA-SON. We can clearly hear the smaller word SON also. The sound spelling focus today is on long **e** spelled **ea**.

Follow the routine next: Body motion, sentence, visualization, and writing.

Write REASON on your whiteboard and add: "eat, treat, seat, sea, please, dream, really, leave, wheat, clean, grease, pleat." For the SON ending, please see the word list in the lesson on PERSON.

Children will make up a sentence they like, and then write and illustrate it. Ex: "The reason I dream I can leave is that I have grease on my pleat."

trouble

Show the children the SnapWord® card for TROUBLE. Study the picture together and ask the children to comment on what they see. Uh oh! There is some trouble in this picture. Apparently a child batted a baseball into a window, breaking it. It looks like he's in trouble! Two sound spellings include **uh** spelled **ou**, and final sound of **l** spelled **le**.

Follow the routine next: Body motion, sentence, visualization, and writing.

Write TROUBLE on your whiteboard and add: "touch, young, double, couple, country, cousin, courage, flourish, nourish, tough, rough, enough."

Children will make up a sentence they like, and then write and illustrate it. Ex: "The trouble is that my young country cousin is not tough enough!"

SNAPWORDS LIST V

List V Words - ABC order:	List V Words by level:
add answer became become begin being believe broke brought build can't care catch caught change complete couldn't died explain feel fell fight finish fix follow forget form grade hear hit kept killed knew learn listen lost mind miss move organize quit rest seen shot speak spend stay stood study succeed talk teach throw travel tried understand wasn't watch win woke worry wouldn't	**Level 1:** add, can't, fell, fix, hit, kept, lost, miss, quit, rest, shot, win **Level 2:** begin, being, broke, feel, forget, grade, hear, killed, mind, seen, stay, study **Level 3:** care, catch, died, form, move, speak, spend, stood, talk, teach, wasn't, watch, woke **Level 4:** became, become, fight, finish, follow, knew, learn, listen, throw, travel, tried, worry **Level 5:** answer, believe, brought, build, caught, change, complete, couldn't, explain, organize, succeed, understand, wouldn't

How To Teach List V, Level 1

add

Show the children the SnapWord® card for ADD. Study the picture together and ask the children to comment on what they see. In this picture, two children who look just alike (**d** and **d**) are going to ADD apples to the basket of apples they already picked. This explanation also conveys the meaning of the math term "add." The only spelling in this word is the final double **d**.

Follow the routine next: Do the body motion from the back of the card, read the sentence, close eyes and do visualization ("see" the word in your mind), and practice writing the word on whiteboards.

Write ADD on your whiteboard. There are many words that end in AD, but not that end in ADD, so we can't really make a word family for this word. Here are some words that begin with ADD: "adding, addict, addend, adder, addle, address."

Children will make up a sentence they like, and then write and illustrate it. Ex: "I will add my apples to yours."

can't

Show the children the SnapWord® card for CAN'T. Study the picture together and ask the children to comment on what they see. This picture shows how contractions are made: some letters get "kicked out" of the word and the apostrophe takes the place of the letters that are now missing. CAN'T is made of CAN and NOT, but when the **n** and the **o** are kicked out, the **t** moves over to stand beside the CAN. The apostrophe is the little black boot that kicked out the letters. Now it is in the spot where the letters used to be.

Follow the routine next: Body motion, sentence, visualization, and writing.

Write CAN'T on your whiteboard and add other "not" contractions: "isn't, aren't, wasn't, weren't, haven't, hasn't, hadn't, won't, wouldn't, don't, doesn't, didn't, couldn't, shouldn't."

Children will make up a sentence they like, and then write and illustrate it. Ex: "I said that I can't and I won't!"

Show the children the SnapWord® card for FELL. Study the picture together and ask the children to comment on what they see. Oh my goodness! One of the girls (the last **l**) just FELL over onto the ground. Do you imagine it is because she was wearing pink high heels in the dirt? Maybe she lost her balance. FELL is a simple word with the only spelling to note is the double **l**.

Follow the routine next: Body motion, sentence, visualization, and writing.

Write FELL on your whiteboard and add: "bell, cell, dell, dwell, jell, knell, sell, tell, well, yell, quell, shell, smell, spell, swell."

** In this section of the Lessons, always write the words in a column so children will be able to see the sound spelling the words have in common.

Sample sentences: "We did yell when it fell." OR "It fell in the cell by the bell."

List V, Level 1

Show the children the SnapWord® card for FIX. Study the picture together and ask the children to comment on what they see. Two children are working together to fix a post (the letter **i**) that is loose and wobbling in its hole. The girl has the post in her hand while the boy holds the dot. I hope this works out!

Follow the routine next: Body motion, sentence, visualization, and writing.

Write FIX on your whiteboard and add: "mix, six, pix, nix, calix, helix, suffix."

Children will make up a sentence they like, and then write and illustrate it. Ex: "I will fix the six pix." OR "Fix the suffix on that word, please."

Show the children the SnapWord® card for HIT. Study the picture together and ask the children to comment on what they see. Two boys are playing **t** ball. It looks like Frank is up to bat. His friend is hiding his eyes. Why do you think he's doing that? Do you think maybe he's standing too close?

Follow the routine next: Body motion, sentence, visualization, and writing.

Write HIT on your whiteboard and add (in a column): "bit, fit, kit, knit, lit, pit, quit, sit, wit, flit, grit, skit, slit, spit, split, twit."

Children will make up a sentence they like, and then write and illustrate it. Ex: "He will hit while I sit." OR "In the skit, I hit the ball into a pit and my bat split!"

kept

Show the children the SnapWord® card for KEPT. Study the picture together and ask the children to comment on what they see. This picture is funny. Harry thinks it is very cool to hold up his arm and his leg. His two friends were amazed at how long he KEPT them up in the air. The only slightly tricky part about this word is that there are two consonants at the end: **p** and **t**. As you sound out KEPT, make sure the children are sounding both consonants, especially as they write the word.

Follow the routine next: Body motion, sentence, visualization, and writing.

Write KEPT on your whiteboard and add in a column: "wept, crept, slept, swept."

Children will make up a sentence they like, and then write and illustrate it. Ex: "He swept while I slept." OR "I kept going as he swept."

List V, Level 1

lost

Show the children the SnapWord® card for LOST. Study the picture together and ask the children to comment on what they see. In this picture Sue is looking for a ring she LOST in the grass. do you think she will find it? LOST is an easy word because you can hear each of the four letters clearly.

Follow the routine next: Body motion, sentence, visualization, and writing.

Write LOST on your whiteboard and add in a column: "cost, frost, accost, defrost, provost."

Children will make up a sentence they like, and then write and illustrate it. Ex: "It cost us when we lost the garden to frost."

miss

Show the children the SnapWord® card for MISS. Study the picture together and ask the children to comment on what they see. In this picture a girl has tears on her cheeks. It seems that her dog went missing and she is telling us that she does MISS him very much. Notice the double **s**. The final **s** is the dog she misses.

Follow the routine next: Body motion, sentence, visualization, and writing.

Write MISS on your whiteboard. Also write: "hiss, kiss, bliss, amiss, dismiss, remiss."

Children will make up a sentence they like, and then write and illustrate it. Ex: "I will not miss the hiss." OR "It is bliss when they dismiss us, but I will miss you!"

quit

Show the children the SnapWord® card for QUIT. Study the picture together and ask the children to comment on what they see. It looks like a couple of kids are in line and the **t** at the end is bumping backwards into people. The **q** is telling him to QUIT doing that! The word is simply the **qu** and then the little word **it**.

Follow the routine next: Body motion, sentence, visualization, and writing.

Write QUIT on your whiteboard and refer to the word list for HIT.

Children will make up a sentence they like, and then write and illustrate it. Ex: "We will quit the skit." OR "The skit was a hit until we quit!"

List V, Level 1

rest

Show the children the SnapWord® card for REST. Study the picture together and ask the children to comment on what they see. This picture makes me relax because the word REST is at rest on the bed! You can hear each of the four sounds in the word pretty easily.

Follow the routine next: Body motion, sentence, visualization, and writing.

Write REST on your whiteboard and add in a column: "best, guest, jest, lest, nest, pest, test, vest, west, zest, blest, chest, crest, quest, wrest."

Children will make up a sentence they like, and then write and illustrate it. Ex: "It will be best if you rest after the test." OR "There is a crest on the chest of my best vest."

shot

Show the children the SnapWord® card for SHOT. Study the picture together and ask the children to comment on what they see. WOW. In this picture you can see a canon being SHOT. The canonball is flying through the air in front of our eyes! The word is easy as it begins with the digraph **sh** and then you can plainly hear the **o** and **t**.

Follow the routine next: Body motion, sentence, visualization, and writing.

Write SHOT on your whiteboard and add: "cot, dot, got, hot, jot, knot, lot, not, pot, rot, tot, blot, clot, plot, slot, spot, trot."

Children will make up a sentence they like, and then write and illustrate it. Ex: "I shot the spot with some water."

win

Show the children the SnapWord® card for WIN. Study the picture together and ask the children to comment on what they see. In this picture, we see an excited cowboy patting his horse and saying, "I am sure my horse will win!" This word is very easy to sound out. It has three small words in it: I, IN, and WIN.

Follow the routine next: Body motion, sentence, visualization, and writing.

Write WIN on your whiteboard and add: "bin, din, fin, gin, kin, pin, sin, tin, chin, grin, shin, skin, spin, thin, twin."

Children will make up a sentence they like, and then write and illustrate it. Ex: "I will grin when I win with my twin!"

List V, Level 2

begin

Show the children the SnapWord® card for BEGIN. Study the picture together and ask the children to comment on what they see. In this picture you can see that the initial **b** is a child with a book. The **i** is another child who is saying, "Please begin to read at the beginning..." This word has no tricky spellings! Identify the small words you can find inside BEGIN. There is I, BE, BEG, IN and GIN.

Follow the routine next: Body motion, sentence, visualization, and writing.

Write BEGIN on your whiteboard and then refer to the word list for WIN which focuses on the IN ending.

** In this section of the Lessons, always write the words in a column so children will be able to see the sound spelling the words have in common.

Children will make up a sentence they like, and then write and illustrate it. Ex: "We will begin to win."

being

Show the children the SnapWord® card for BEING. Study the picture together and ask the children to comment on what they see. Oh my! In this picture, the letter **i** is a child who is obviously being obnoxious. His dad looks surprised or maybe even freaked out! I wonder what Henry (the child) is screaming about? This word is super easy. Just the word BE with the **ing** ending.

Follow the routine next: Body motion, sentence, visualization, and writing.

Write BEING on your whiteboard and add words that you can add ING to. For example: see (seeing), run (running)*, cook (cooking), sweep (sweeping), paint (painting). Let the students help you come up with a list of verbs, then add ING together. See the notes below about adding ING.

Children will make up a sentence they like, and then write and illustrate it. Ex: "Henry is being loud!" Note that an adjective will follow the word BEING. The adjective will tell you what they are being. Being loud? Being slow?

*For words with a short vowel followed by a consonant, double the consonant. This will keep the vowel short.

**For words ending in an E, remove the E before adding ING.

List V, Level 2

broke

    Show the children the SnapWord® card for BROKE. Study the picture together and ask the children to comment on what they see. Uh oh! The family is shocked to see that the **o** in the center of BROKE...well, broke. The sound spelling in this word is **o-e** (long **o** with Pinchy **e**).
    Follow the routine next: Body motion, sentence, visualization, and writing.
    Write BROKE on your whiteboard and add: "joke, poke, woke, yoke, choke, smoke, spoke, stoke, stroke, awoke, evoke, invoke, revoke."
    Children will make up a sentence they like, and then write and illustrate it. Ex: "He broke the spoke with one stroke." "When it broke, he awoke!" "It was no joke that it broke."

feel

    Show the children the SnapWord® card for FEEL. Study the picture together and ask the children to comment on what they see. In this picture we can see that the boy who is the **f** doesn't FEEL well! His friend, the **l** is handing him a tissue while he coughs. The target sound spelling, **ee**, forms a table that is holding the box of tissue.
    Follow the routine next: Body motion, sentence, visualization, and writing.
    Write FEEL on your whiteboard and add: "beet, beef, beep, bleed, bleep, breed, cheek, creed, creek, creep, greed, green, Greek, need, weed, greet, leech, preen, screech, screen, sheen, sheep, speech, sleep, speed, steed, teeth, tweet."
    Children will make up a sentence they like, and then write and illustrate it. Ex: "I feel I'd like to have beef and beets for lunch." "The sheep will bleet if they feel the need."

forget

    Show the children the SnapWord® card for FORGET. Study the picture together and ask the children to comment on what they see. This is an interesting picture. The two round letters, **o** and **g** are fishbowls. The **f** is Dad who is saying, "Don't forget to feed the fish" to his son, the **r**. This is a compound word made of two easy, small words: FOR and GET.
    Follow the routine next: Body motion, sentence, visualization, and writing.
    Write FORGET on your whiteboard and add: "diet, duet, poet, whet, asset, beget, beset, cadet, comet, covet, egret, inlet, inset, facet."
    Children will make up a sentence they like, and then write and illustrate it. Ex: "The cadet will forget his diet." "I will never forget the comet in the inlet."

List V, Level 2

grade

Show the children the SnapWord® card for GRADE. Study the picture together and ask the children to comment on what they see. This picture shows two meanings for the word GRADE. It is a first GRADE class and Ms. Swift is going to GRADE papers now. The target sound spelling for this word is long **a** spelled **a-e** (**a** with Pinchy **e**).

Follow the routine next: Body motion, sentence, visualization, and writing.

Write GRADE on your whiteboard and add: "blade, evade, glade, shade, spade, trade, arcade, decade, invade, parade, fade, wade, jade, lade, made."

Children will make up a sentence they like, and then write and illustrate it. Ex: "My grade will see the parade in the shade." "I made the grade!"

hear

Show the children the SnapWord® card for HEAR. Study the picture together and ask the children to comment on what they see. I love this picture that shows the boy and girl saying, "I can HEAR a bluebird singing." The sound spelling to pay attention to is EAR, which is also a small word inside of HEAR.

Follow the routine next: Body motion, sentence, visualization, and writing.

Write HEAR on your whiteboard and add: "dear, fear, year, lear, near, rear, tear, year, clear, drear, smear, spear, appear, endear."

Children will make up a sentence they like, and then write and illustrate it. Ex: "He can hear with his ear." "I fear we are near the rear."

killed

Show the children the SnapWord® card for KILLED. Study the picture together and ask the children to comment on what they see. Oh my goodness! The lightning bolts hit two pine trees and killed them! I would like to tell the boy who is pointing this out to us to go inside now! It is dangerous to be outside during a thunder storm! The spellings to note are the double **l** in Kill and the final **ed** to show past tense.

Follow the routine next: Body motion, sentence, visualization, and writing.

Write KILLED on your whiteboard and add: "billed, filled, milled, spilled, grilled, trilled, swilled, thrilled, willed, chilled, drilled."

Children will make up a sentence they like, and then write and illustrate it. Ex: "He milled and drilled the trees that were killed in the storm."

List V, Level 2

mind

Show the children the SnapWord® card for MIND. Study the picture together and ask the children to comment on what they see. This picture shows what I assume is a very smart person who has a good MIND! The word MIND is used in another way also: If you don't MIND something, that means it's ok. You don't love it, but you don't mind it either. This child is saying, "I don't mind spaghetti." Ask the students for examples of things they don't mind. For example: I don't mind when it rains because I have a garden that needs water. The sound spelling is long **i** spelled with an **i** by itself.

Follow the routine next: Body motion, sentence, visualization, and writing.

Write MIND on your whiteboard and add: "bind, find, hind, kind, rind, wind (long i), blind, grind, behind."

Children will make up a sentence they like, and then write and illustrate it. Ex: "I do mind if I am behind!" "I find I mind if you are not kind!"

seen

Show the children the SnapWord® card for SEEN. Study the picture together and ask the children to comment on what they see. In this picture, the missing dog is the **s**. He is behind a tree and his owner can't see him. She is asking, "Have you SEEN my lost dog?" The sound spelling in this word is long **e** spelled **ee**. We practiced this sound spelling when we studied the word FEEL.

Follow the routine next: Body motion, sentence, visualization, and writing.

Write SEEN on your whiteboard and add: "keen, queen, teen, green, preen screen, sheen." Also refer to the word list for FEEL.

Children will make up a sentence they like, and then write and illustrate it. Ex: "I have seen a queen by the green screen!" Please be sure to prompt the children to always use the word "have" before the word SEEN. It is always, "I have seen..." It is never, "I seen..." "I have seen trees fall." "I have seen that dog before." "I have seen five red cars today." "I have seen you before!"

stay

Show the children the SnapWord® card for STAY. Study the picture together and ask the children to comment on what they see. In this picture, Mom, who is the **t**, is telling her dog (the **s**) to STAY. It looks like Mom is going shopping and wants the dog to STAY behind and not follow her. The sound spelling in this word is long **a** spelled **ay**.

Follow the routine next: Body motion, sentence, visualization, and writing.

Write STAY on your whiteboard and add: "bay, day, hay, jay, lay, may, nay, pay, quay, ray, say, way, bray, clay, cray, ray, gray, play, pray, slay, spray, stay, stray, sway, tray."

Children will make up a sentence they like, and then write and illustrate it. Ex: "I can stay all day in the bay." "Let's stay and play in the spray all day."

List V, Level 2
study

Show the children the SnapWord® card for STUDY. Study the picture together and ask the children to comment on what they see. Wow these two girls are really studying hard. The tall letters **t** and **d** are the girls, while the **u** and the **y** hold their books. This word is simple: you can hear each sound. The only tricky thing is the final **y** who is pretending to be an **e**. Remember that in one syllable words, **y** pretends to be Long **i**, but in words with more than one syllable, the final **y** pretends to be Long **e**.

Follow the routine next: Body motion, sentence, visualization, and writing.

Write STUDY on your whiteboard and add: "baldy, bandy, beady, buddy, candy, caddy, daddy, dandy, giddy, moldy, needy, rowdy, shady, tardy, weedy, windy, wordy."

Children will make up a sentence they like, and then write and illustrate it. Ex: "I study with a buddy." "I will study under a shady tree."

List V, Level 3

care

Show the children the SnapWord® card for CARE. Study the picture together and ask the children to comment on what they see. In this picture we see three people. The **c** and the **e** persons are both looking away from the girl in the middle who is the **r**. She apparently has lost her dog and is mentioning that the others don't CARE that her dog ran away. The sound spelling for CARE is **are** that sounds like **air**. It is important to show students many cases of -ARE words; they need to be familiar with this spelling pattern as it cannot be sounded out.

Follow the routine next: Body motion, sentence, visualization, and writing.

Write CARE on your whiteboard and add: "bare, dare, fare, hare, mare, pare, rare, tare, ware, glare, scare, share, snare, spare, square, stare, aware, beware."

** In this section of the Lessons, always write the words in a column so children will be able to see the sound spelling the words have in common.

Children will make up a sentence they like, and then write and illustrate it. Ex: "The hare will care if he's bare." "The mare does care for the hare."

catch

Show the children the SnapWord® card for CATCH. Study the picture together and ask the children to comment on what they see. In this picture, the **t** is a catcher who is about to CATCH the baseball. The **t** stands out in this picture because it is really not heard when you say CATCH. We pronounce this word like CACH. The girl who is the **c** is ducking so the ball won't hit her. Children can remember the **t** in CATCH also by noting the smaller word CAT. Say, "I want to catch the cat" by way of helping the children remember this spelling.

Follow the routine next: Body motion, sentence, visualization, and writing.

Write CATCH on your whiteboard and add: "batch, botch, ditch, etch, fetch, hatch, hitch, hutch, itch, latch, match, notch, patch, pitch, retch, witch, stitch, crutch, glitch."

Children will make up a sentence they like, and then write and illustrate it. Ex: "At the match, there was a hitch with the catch." "I will catch the batch of cookies."

died

Show the children the SnapWord® card for DIED. Study the picture together and ask the children to comment on what they see. In this image, three kids are working in their garden. The child who is the **i** is saying, "Our garden DIED when we forgot to water it." The focus on this word is the **ied** ending. The root word is "die" and to make it past tense, you just add a **d**. Other words that end in **ied**, however, are root words that end in **y** (dry, fry, try, pry) and when you make these words past tense, you have to change the **y** to an **i** first and then add **ed**. Otherwise you would have "dryed, fryed, tryed, pryed" which would be two-syllable words. When you change the **y** to an **i** and add **ed**, the **ie** makes the long **i** spelling.

Follow the routine next: Body motion, sentence, visualization, and writing.

Write DIED on your whiteboard and add: "lied, dried, fried, tried, pried, relied."

Children will make up a sentence they like, and then write and illustrate it. Ex: "The plants first dried up, then died."

List V, Level 3

form

Show the children the SnapWord® card for FORM. Study the picture together and ask the children to comment on what they see. In this picture, two boys FORM clay into a big gawky turtle. It actually looks fun, doesn't it? Another meaning for FORM is a paper with blanks you fill out. The focus sound spelling is the **or**, which is also a small word inside the larger word.

Follow the routine next: Body motion, sentence, visualization, and writing.

Write FORM on your whiteboard and add: "corm, norm, dorm, storm, inform, deform, conform, perform, uniform."

Children will make up a sentence they like, and then write and illustrate it. Ex: "The form about the storm is in the dorm."

move

Show the children the SnapWord® card for MOVE. Study the picture together and ask the children to comment on what they see. This picture shows a truck that you use to MOVE big things. It looks like it is moving quickly down the street. This word MOVE is tricky in terms of its spelling and its sounds. The **ove** looks like it is a Long **o** with Pinchy **e** spelling, but in reality it sounds like "**oove**".

Follow the routine next: Body motion, sentence, visualization, and writing.

Write MOVE on your whiteboard and add: "prove, remove, reprove, approve, improve, disprove." Note that all these words have either MOVE or PROVE as their root.

Children will make up a sentence they like, and then write and illustrate it. Ex: "We will remove the trash and move in soon."

speak

Show the children the SnapWord® card for SPEAK. Study the picture together and ask the children to comment on what they see. This picture shows a lady watching a man SPEAK. She apparently loves what he's saying. There is a table holding water glasses made of the sound spelling **ea** (long **e**), which is our focus for this word. All other sounds are simple and can easily be heard.

Follow the routine next: Body motion, sentence, visualization, and writing.

Write SPEAK on your whiteboard and add: "beak, weak, leak, peak, teak, freak, tweak, bleak, sneak, creak." There are many words with the target sound spelling EA, but the ones listed here are the word family EAK.

Children will make up a sentence they like, and then write and illustrate it. Ex: "I will peak when he begins to speak."

List V, Level 3

spend

Show the children the SnapWord® card for SPEND. Study the picture together and ask the children to comment on what they see. In this picture, a lady is looking at a Teddy bear and is trying to decide how much she wants to SPEND on it. SPEND is an easy word because you can hear all 5 sounds, and each sound is just one letter. Look for smaller words: "PEN, END, PEND."

Follow the routine next: Body motion, sentence, visualization, and writing.

Write SPEND on your whiteboard and add: "bend, pend, vend, fend, rend, wend, lend, send, mend, tend, amend, blend, trend, extend, defend, legend, depend, offend, attend."

Children will make up a sentence they like, and then write and illustrate it. Ex: "I will spend a lot to mend and tend the fence."

stood

Show the children the SnapWord® card for STOOD. Study the picture together and ask the children to comment on what they see. Wow. In this picture five people stood on the edge of a cliff and looked down at a flowing river far below. The sound spelling for STOOD is short **oo** as in BOOK.

Follow the routine next: Body motion, sentence, visualization, and writing.

Write STOOD on your whiteboard and add: "book, brook, cook, crook, foot, good, hood, hoof, hook, look, nook, rook, shook, soot, took, wool, cookie."

Children will make up a sentence they like, and then write and illustrate it. Ex: "Cookie stood by the brook with her book."

Show the children the SnapWord® card for TALK. Study the picture together and ask the children to comment on what they see. This picture is funny, but also helpful. Apparently the **l** is falling into a hole, which draws attention to this letter we don't really hear. The parents are saying, "We need to talk about what to do with the **l**". The sound spelling of note here is **al** which sounds like **ah**; the same sound as in "water, wash, watch, father."

Follow the routine next: Body motion, sentence, visualization, and writing.

Write TALK on your whiteboard and add: "walk, stalk, chalk."

Children will make up a sentence they like, and then write and illustrate it. Ex: "I can walk and talk with my stalk and my chalk."

List V, Level 3

teach

Show the children the SnapWord® card for TEACH. Study the picture together and ask the children to comment on what they see. In this picture, Dad will TEACH his son how to fish. Right now, however, the son looks a bit nervous about the fishing pole swishing right at his nose! The focus sound spelling, again, is long **e** spelled **ea**. Refer also to the lesson on SPEAK.

Follow the routine next: Body motion, sentence, visualization, and writing.

Write TEACH on your whiteboard and add these words from the word family "each": "each, beach, leach, peach, reach, bleach, breach, preach."

Children will make up a sentence they like, and then write and illustrate it. Ex: "I can teach you how to reach for the peach."

wasn't

Show the children the SnapWord® card for WASN'T. Study the picture together and ask the children to comment on what they see. This picture shows the action of making a contraction; the **o** is being kicked out of the word "NOT" and is being replaced by the black boot. Also refer to the story for CAN'T in Level 1 of this resource. This word is made up of WAS and NOT, but with the **o** missing.

Follow the routine next: Body motion, sentence, visualization, and writing.

Write WASN'T on your whiteboard and add "not contractions" from the lesson on CAN'T.

Children will make up a sentence they like, and then write and illustrate it. Ex: "It wasn't fun to have to do what I can't do well."

watch

Show the children the SnapWord® card for WATCH. Study the picture together and ask the children to comment on what they see. This picture shows two guys wearing ticking watches. The boy in the middle is saying, "Both guys wanted me to WATCH their watches." That is a bit like watching paint dry! Please refer to lesson on CATCH which has the same target spelling.

Follow the routine next: Body motion, sentence, visualization, and writing.

Write WATCH on your whiteboard and add the word list from CATCH lesson.

Children will make up a sentence they like, and then write and illustrate it. Ex: "I will watch you catch the ball."

List V, Level 3

woke

Show the children the SnapWord® card for WOKE. Study the picture together and ask the children to comment on what they see. The picture for WOKE almost tells a whole story. We can see a mother holding her crying baby while a rooster crows loudly to one side. She is saying, "The rooster woke the baby!" The sound spelling focus is on long **o** spelled **o-e**.

Follow the routine next: Body motion, sentence, visualization, and writing.

Write WOKE on your whiteboard and add: "coke, poke, yoke, joke, awoke, evoke, block, spoke, smoke, broke, choke, stoke, stroke, revoke, invoke, provoke."

Children will make up a sentence they like, and then write and illustrate it. Ex: "I woke to the smell of smoke." "I woke her with a little poke."

List V, Level 4

became

Show the children the SnapWord® card for BECAME. Study the picture together and ask the children to comment on what they see. The boy who is the **b** at the beginning of the word is happy to see that the caterpillar became a butterfly. Notice two other lower case **e** caterpillars that have not transformed yet. The word BECAME is made of two smaller words. BECAME is a past tense word meaning the action of turning into a butterfly already happened. The target sound spelling is long **a** spelled **a-e** (Pinchy **e**).

Follow the routine next: Body motion, sentence, visualization, and writing.

Write BECAME on your whiteboard and add: "same, lame, tame, name, game, shame, blame, flame, frame."

Children will make up a sentence they like, and then write and illustrate it. Ex: "To our shame, the name of the game became lame!"

become

Show the children the SnapWord® card for BECOME. Study the picture together and ask the children to comment on what they see. This word actually happens before BECAME. In this picture we can see both the boy and the caterpillar waiting and dreaming of the caterpillar transforming into a butterfly. It hasn't happened yet! Again, BECOME is made of two words and the sound spelling is **o-e** when it sounds like **uh**.

Follow the routine next: Body motion, sentence, visualization, and writing.

Write BECOME on your whiteboard and add other words where **o-e** sound like **uh**: "some, come, dove, love, shove, above, dove."

Children will make up a sentence they like, and then write and illustrate it. Ex: "Some baby doves I love will become grown up doves!"

fight

Show the children the SnapWord® card for FIGHT. Study the picture together and ask the children to comment on what they see. Oh my! This picture shows a very active snowball fight. The **g** is a girl who fell back into the snow and the **h** is a girl who just got a snowball in the face! The sound spelling in FIGHT is long **i** spelled **igh**.

Follow the routine next: Body motion, sentence, visualization, and writing.

Write FIGHT on your whiteboard and add "sight, might, night, knight, right, tight, blight, bright, flight, fright, plight, slight."

Children will make up a sentence they like, and then write and illustrate it. Ex: "What a sight to see the knight in a fight at night."

List V, Level 4

finish

Show the children the SnapWord® card for FINISH. Study the picture together and ask the children to comment on what they see. The two **i**'s are kids singing. You can find some small words inside FINISH: FIN, IN, IS. The only sound spelling of note is the digraph **sh**.

Follow the routine next: Body motion, sentence, visualization, and writing.

Write FINISH on your whiteboard and add: "dish, fish, wish, swish, squish."

Children will make up a sentence they like, and then write and illustrate it. Ex: "I wish you would finish the fish on your dish."

follow

Show the children the SnapWord® card for FOLLOW. Study the picture together and ask the children to comment on what they see. In FOLLOW, the three tall letters are confused boys who are not sure who they should follow. The boy who is the **f** is saying, "Follow me!" But the other boys are distracted by the barking turtle and the quacking duck. What do you think will happen? Sound spellings are the double **l** and the long **o** spelled **ow**.

Follow the routine next: Body motion, sentence, visualization, and writing.

Write FOLLOW on your whiteboard and add: "low, know, mow, row, sow, tow, blow, crow, flow, glow, grow, show, slow, snow, stow, throw."

Children will make up a sentence they like, and then write and illustrate it. Ex: "I know to follow the slow crow into the snow."

knew

Show the children the SnapWord® card for KNEW. Study the picture together and ask the children to comment on what they see. In this picture, the silent **k** is a boy who is showing the way as the others followed him. Other silent **k** words include "know, knife, knee, knapsack, knight." The other sound spelling is **ew**.

Follow the routine next: Body motion, sentence, visualization, and writing.

Write KNEW on your whiteboard and add: "dew, few, hew, new, pew, blew, brew, chew, crew, drew, flew, screw, stew, threw."

Children will make up a sentence they like, and then write and illustrate it. Ex: "I knew that a few of the new crew blew on the stew."

List V, Level 4

learn

Show the children the SnapWord® card for LEARN. Study the picture together and ask the children to comment on what they see. Wow, I see a very complicated mathematical equation on the chalkboard! The student is nodding his head saying he is sure he can learn it! The sound spelling is **ear** that sounds like **er**.

Follow the routine next: Body motion, sentence, visualization, and writing.

Write LEARN on your whiteboard and add: "earn, earth, early, heard, earl, pearl, search."

Children will make up a sentence they like, and then write and illustrate it. Ex: "The earl did learn to search the earth for pearls."

listen

Show the children the SnapWord® card for LISTEN. Study the picture together and ask the children to comment on what they see. In this picture, Mom is giving instructions and her son is listening to her. He must agree with what she is saying because he is nodding his head. We emphasize the **t** in this word by making him into a child because it is not a letter we can hear. The sound spelling in this word is the sound of **s** spelled **st**.

Follow the routine next: Body motion, sentence, visualization, and writing.

Write LISTEN on your whiteboard and add: "castle, rustle, whistle, bustle."

Children will make up a sentence they like, and then write and illustrate it. Ex: "In the castle, I listen to leaves rustle and the wind whistle."

throw

Show the children the SnapWord® card for THROW. Study the picture together and ask the children to comment on what they see. Someone's about to be dunked in the pool! THROW is made of the digraph **th** and the little word ROW. The sound spelling is the one on which we focused on p. 109, in the lesson on FOLLOW - long **o** spelled **ow**.

Follow the routine next: Body motion, sentence, visualization, and writing.

Write THROW on your whiteboard and refer to the word list from the lesson on FOLLOW.

Children will make up a sentence they like, and then write and illustrate it. Ex: "In the winter, I do throw snowballs down low."

List V, Level 4

travel

Show the children the SnapWord® card for TRAVEL. Study the picture together and ask the children to comment on what they see. This family is all packed up and going on a trip! Notice which letters are people and which are bags. The sound spelling for TRAVEL is **el** that sounds like **l**.

Follow the routine next: Body motion, sentence, visualization, and writing.

Write TRAVEL on your whiteboard and add: "angel, counsel, panel, marvel, label, parcel."

Children will make up a sentence they like, and then write and illustrate it. Ex: "When I travel, I label my parcel on the side panel."

tried

Show the children the SnapWord® card for TRIED. Study the picture together and ask the children to comment on what they see. Ouch! It looks like our boy was lifting weights that are too heavy for him! He TRIED very hard though! The sound spelling here is **ied** (final **y** turns into **ied** to make past tense).

Follow the routine next: Body motion, sentence, visualization, and writing.

Write TRIED on your whiteboard and add: "cried, fried, dried, pried, plied, spied."

Children will make up a sentence they like, and then write and illustrate it. Ex: "He tried the fried food he spied."

worry

Show the children the SnapWord® card for WORRY. Study the picture together and ask the children to comment on what they see. In this picture two kids worry about Humpty Dumpty who fell from the wall. But don't WORRY - they are promising to fix him up! Sound spellings are the double **r** and the final **y** that sounds like an **e**. The **y** always sounds like an **e** at the end of words with more than one syllable. There is also the sound spelling **or** that sounds like **er**.

Follow the routine next: Body motion, sentence, visualization, and writing.

Write WORRY on your whiteboard and add: "blurry, hurry, scurry, flurry, slurry."

Children will make up a sentence they like, and then write and illustrate it. Ex: "Don't worry. I willhurry and scurry in the flurry of snow."

List V, Level 5

answer

Show the children the SnapWord® card for ANSWER. Study the picture together and ask the children to comment on what they see. Apparently, someone over the hill is yelling for help. Three kids are hustling to find him and are yelling, "we're coming!" They are going to ANSWER his call for help! The odd spelling in this word is the **sw** that sounds like **s**. The final **er** is another sound spelling, but it is very common.

Follow the routine next: Body motion, sentence, visualization, and writing.

Write ANSWER on your whiteboard and add: "sword," the only other word I know of in which /S/ is spelled SW.

Children will make up a sentence they like, and then write and illustrate it. Ex: "The answer to this fight is NOT a sword!"

believe

Show the children the SnapWord® card for BELIEVE. Study the picture together and ask the children to comment on what they see. Two people can't BELIEVE how many **e**'s are in the word BELIEVE! When you analize the word, notice the BE at the beginning, then you can see LIE (pronounced LEE) and then the final **ve** ending. Also EVE is at the end of the word. The focus for sound spellings will be **ie** that sounds like long **e**.

Follow the routine next: Body motion, sentence, visualization, and writing.

Write BELIEVE on your whiteboard and add: "thief, grief, relief, chief, niece, fierce, achieve, shriek, yield, shield."

Children will make up a sentence they like, and then write and illustrate it. Ex: "The niece of the chief did shriek when she saw the fierce theif!"

brought

Show the children the SnapWord® card for BROUGHT. Study the picture together and ask the children to comment on what they see. I love this picture because it simplifies the word BROUGHT so much! You can see **b** and **r** at the beginning, then all the letters over the balloon say short **o** together (**ough**). Finally, a simple **t** sound brings up the rear. You sound this word like this: B-R-OUGH-T. The sound spelling of course is **ough** that sounds like short **o**.

Follow the routine next: Body motion, sentence, visualization, and writing.

Write BROUGHT on your whiteboard and add: "ought, bought, fought, thought."

Children will make up a sentence they like, and then write and illustrate it. Ex: "I thought they ought to have brought the candy they bought and not have fought for more."

List V, Level 5

build

Show the children the SnapWord® card for BUILD. Study the picture together and ask the children to comment on what they see. Two construction workers are the rather unusual spelling for short I. Notice that the I is the one doing the speaking (just like in the word build in which the **u** is just lending moral support to the **i**). The sentence, "You and I can BUILD this by ourselves" helps to remind students of the way short I is spelled (You and I - U and I).
Follow the routine next: Body motion, sentence, visualization, and writing.
Write BUILD on your whiteboard and add: "built."
Children will make up a sentence they like, and then write and illustrate it. Ex: "I can build what you built!"

caught

Show the children the SnapWord® card for CAUGHT. Study the picture together and ask the children to comment on what they see. This is a rather complicated-looking word, but the picture simplifies it a whole lot. You can see the **c** at the beginning and the **t** at the end. In the middle are speckled letters that all work together to make the short **o** sound. In order to remember this combination of letters, notice the pattern of rounded letter, "cup" letter, rounded letter, upside down "cup" letter. Our sound spelling is **augh** sounding like short **o**.
Follow the routine next: Body motion, sentence, visualization, and writing.
Write CAUGHT on your whiteboard and add: "taught, daughter, naughty."
Children will make up a sentence they like, and then write and illustrate it. Ex: "I caught and taught my naughty daughter." (Sounds like she ran away from class and I had to run out and find her.)

change

Show the children the SnapWord® card for CHANGE. Study the picture together and ask the children to comment on what they see. This picture is all about making change for paper money. The word starts with the digraph **ch** (the person waving the dollar bill), next comes small word AN, and finally **ge** sounding like **j**. The **ge** is digging in her purse trying to find CHANGE - and much of it is scattered on the floor now!
Follow the routine next: Body motion, sentence, visualization, and writing.
Write CHANGE on your whiteboard and add: "range, strange, arrange, mange, grange."
Children will make up a sentence they like, and then write and illustrate it. Ex: "I can arrange to make change by the range."

List V, Level 5

complete

Show the children the SnapWord® card for COMPLETE. Study the picture together and ask the children to comment on what they see. In this picture we see our builders again. It looks like they might have hit a snag in their work as one of them is saying, "We must COMPLETE this job!" **m** worker is looking discouraged or mad, while **t** worker is studying the building they need to complete. Our focus is on the ETE ending. The word begins with what sounds like the word COME, except that it is missing the final E. Also, the **o** sounds like short **u**. The **ete** ending sounds like EAT.

Follow the routine next: Body motion, sentence, visualization, and writing.

Write COMPLETE on your whiteboard and add: "replete, delete, concrete, compete, delete, athlete, discrete, excrete."

Children will make up a sentence they like, and then write and illustrate it. Ex: "We will compete in making concrete to complete this job."

couldn't

Show the children the SnapWord® card for COULDN'T. Study the picture together and ask the children to comment on what they see. The word COULDN'T is made of two words - sort of. COULD and NOT - only NOT is missing it's **o**. In our story of contractions remember that the black boot sometimes randomly kicks letters out of word. The rest of the pieces huddle together, making a brand new word. So we go from COULD NOT to COULDN'T. The sound spelling is **oul** sounding like the **oo** in book.

Follow the routine next: Body motion, sentence, visualization, and writing.

Write COULDN'T on your whiteboard and add: "could, would, should, wouldn't, shouldn't."

Children will make up a sentence they like, and then write and illustrate it. Ex: "We couldn't believe she would kick the O."

explain

Show the children the SnapWord® card for EXPLAIN. Study the picture together and ask the children to comment on what they see. Oh my. Someone has been playing with matches - and now he's trying to EXPLAIN to Dad why he was playing with matches. our spelling focus is on the **ain** ending. This is not a complicated spelling. It is like **ai** in rain.

Follow the routine next: Body motion, sentence, visualization, and writing.

Write EXPLAIN on your whiteboard and add: "train, maintain, retain, strain, brain, obtain, remain, regain, distain, contain, abstain, pertain, stain, sprain."

Children will make up a sentence they like, and then write and illustrate it. Ex: "I can explain why I can't contain the fire on the train."

List V, Level 5

organize

Show the children the SnapWord® card for ORGANIZE. Study the picture together and ask the children to comment on what they see. In this picture we see three kids, three jars, and a whole lot of marbles on the floor. Each syllable in the word OR-GAN-IZE has a person with it. Breaking a long word up into its syllables really helps with reading and writing the word! The sound spellings in this word are the familiar **or**, and the long **i** ending IZE. We will focus on words that end in **ize**.

Follow the routine next: Body motion, sentence, visualization, and writing.

Write ORGANIZE on your whiteboard and add: "ionize, resize, realize, itemize, outsize, stylize, utilize, vitalize, theorize, notarize, harmonize, equalize, colonize."

Children will make up a sentence they like, and then write and illustrate it. Ex: "We will utilize these jars to organize our outsize set of marbles." Or, "We realize we need to organize our marbles."

succeed

Show the children the SnapWord® card for SUCCEED. Study the picture together and ask the children to comment on what they see. This picture is helpful in remembering the two sets of double letters as the two **c**'s are dressed the same and so are the two **e**'s. The kids are commenting that they didn't succeed in being all the same height! Note the one tall letter at the end of the word. Our focus for this word will be on **eed** ending, but also on the **cc** spelling that sounds like **ks**. We will have two word lists; one for each spelling.

Follow the routine next: Body motion, sentence, visualization, and writing.

Write SUCCEED on your whiteboard and add EED words: "leveed, agreed, exceed, reseed, indeed, ragweed." Now add some words with CC that sounds like KS: success, access, accept, vaccine, accent, accessory, accelerate."

Children will make up a sentence they like, and then write and illustrate it. Ex: "We agreed we will succeed indeed!" And, "I accept the success of our vaccine."

understand

Show the children the SnapWord® card for UNDERSTAND. Study the picture together and ask the children to comment on what they see. This picture is helpfully pointing out that UNDERSTAND is a compound word. The UNDER part is under the house, while the STAND part is demonstrated by two standing men. They together don't UNDERSTAND how to get inside the house. There are no stairs, you see. There is no tricky sound spelling for this word - **er** is familiar by now. So we will do AND word family.

Follow the routine next: Body motion, sentence, visualization, and writing.

Write UNDERSTAND on your whiteboard and add: "bland, brand, grand, gland, stand, expand, unhand, errand, island, disband, remand, inland."

Children will make up a sentence they like, and then write and illustrate it. Ex: "I understand the brand needs to expand on the island!"

wouldn't

Show the children the SnapWord® card for WOULDN'T. Study the picture together and ask the children to comment on what they see. This picture is very much like the picture for COULDN'T. In this picture, again, the black boot has kicked the **o** out of NOT. The people are upset and said, "We WOULDN'T want this to happen to us!" The target sound spelling is colored blue so it is easy to identify. **oul** sounding like **oo** as in BOOK.

Follow the routine next: Body motion, sentence, visualization, and writing.

Write WOULDN'T on your whiteboard and add: "could, would, should, couldn't, shouldn't"

Children will make up a sentence they like, and then write and illustrate it. Ex: "I wouldn't want to lose my O. She shouldn't have done that!"

HOW TO TEACH
SnapWords®
high-frequency picture words

activities and games that will help you successfully help children recognize words, comprehend their meaning, use them to make phrases and sentences, and use the SnapWords® as writing prompts

by Sarah Major, M.Ed.

ABOUT PART 2

How to Teach SnapWords® is comprised of activities and games that will help you successfully teach children to instantly recognize their sight words using stylized SnapWords®. But learning to recognize words is just the beginning! Children will go on from there to using the high-frequency words to make phrases and sentences. Throughout the activities in this little book, the focus is on the meaning and correct usage of the words, using alternate ways of learning that allow children to avoid memorization and drill. The color, humor, and body motions utilized attract children to the process of learning to read and use words.

To begin, choose a group of words to display in a pocket chart. Before you begin to play the games, tell the children what each word says before they have a chance to guess, and possibly guess incorrectly. Once a child has absorbed the image hearing himself say the wrong word, that incorrect word will become stuck in his memory and it will be difficult later to replace that first impression with the right word. So, avoid the issue by simply telling the children what each word says.

The activities in *How to Teach SnapWords®* fall into five main categories from simple word recognition, to reading activities such as making phrases and sentences, to activities that link reading and writing, and finally to studying the structure of words through spelling patterns.

For each of the activities, focus on enjoyment rather than drill and tedium. Follow the children's focus, and when their attention seems to begin to lag, wind it up for the day and move on to something else. The images are powerful vehicles for learning, and children will learn the words surprisingly quickly!

esol considerations

- This book of games and activities is vital tool for ESOL settings.
- The visuals provide powerful meaning-makers for English language learners of any age.
- The cards become visual prompts to the meaning of English words and a visual that ties meaning to print.
- Putting sentences together guides the English Language Learner into an understanding of English sentence structure and word usage.
- Motions also aid in the understanding of the meaning of the words and phrases.

WORD RECOGNITION - learning

Introducing SnapWords®

- Use SnapWords® Teaching Cards.
- Choose 10-12 cards to display in a pocket chart.
- Tell the child(ren) what each word says.
- Talk about what you see in the pictures.
- Do the body motion found on the reverse side.
- Use the word in the sentence from the reverse side.
- After introducing the words, go back over the set, saying each word together. When the children see the words and images, do the body motions, and hear themselves saying the words, learning will be strengthened.

pop up

- Use the same SnapWords® Teaching Cards from game 1.
- With the class near you, explain that you are going to play a game in which they will pop up when they hear their name and will come up to the pocket chart to point to a word they can recognize.
- Encourage children to select a word when it is not their turn so they are ready when their name is called.
- Do not use this activity for teaching; rather keep it moving quickly so that no child gets bored. Stop the game as soon as interest begins to wander.

WORD RECOGNITION - learning

where's word-o?

3

• Use the same set of SnapWords® Teaching Cards.
• Have a fun pointing stick.
• Write children's names on popsicle sticks.
• Tell the children when they hear their name, they will come get the pointer and quickly locate the word you say.
• Choose your first name by selecting a stick with a name on it, and ask, "Where's PLAY, Jaylen?"
• Jaylen will jump up, grab the pointing stick and locate PLAY.
• Call names until everyone has had at least one turn, keeping the play moving quickly.
• If a child selects the wrong word, point out the beginning sound and keep on going. Make a quick note of the missed words.

which is which?

4

• Use the same SnapWords® Teaching Cards.
• Use the name sticks again in order to be sure all play.
• Choose your first name.
• Select two cards and hold them up asking, "Natalie, which word says PLAY?"
• Put the words back into the pocket chart and start again.
  With larger groups, you can have teams that take turns.

WORD RECOGNITION - review & practice

 around the world

• Use your set of popsicle sticks.
• Tell the children you are going to play a game and you will start it in order to show them how it is played.
• Select the first child and point to a word.
• That child will read the word, then point to another word for a new child to read. You can either choose sticks or let each child choose a name for himself.
• Continue until everyone has had a turn or two and all the words are thoroughly reviewed.

war, played in threes

• One person is the word caller.
• The word caller will display the first word.
• The child who correctly reads the word first gets the card.
• Go through all the words and then change the name caller to be another child.
• Continue until all three children have been the name caller.

 group games

Use the same words, but this time, print them on plain cards. You will need 4 copies of each word.

GO FISH

• The goal of this game is to make as many matches as you can.
• Use 4 copies of each word and shuffle well.
• Deal 5 cards to each player, face down.
• Each child will look at their hand first to see if they have any matching cards they can lay on the table in front of them.
• Take turns going around the circle asking "Jason, do you have the word PLAY?" If Jason does have the word, he gives it to the requestor. If he doesn't have the word, he will reply, "Go fish" and the first player will draw a card from the pile.

MAKING MATCHES
• Use two sets of the printed cards, shuffled and laid out face up on the table.
• Children will take turns finding two words that match and reading them for their partner.

MEMORY
• Use two sets of the printed cards, shuffled and laid out face down on the table.
• Children will take turns turning over two cards to see if they can make a match. If they can, they keep the cards.

WORD RECOGNITION - review & assessment

whole group review or quiz

• Use your SnapWords® groups of cards, but display them with the backs of the cards showing.
• Children will have either whiteboards and markers, or pencil and paper, depending on whether you want to review or quiz them on the words.
• Without pointing to it, choose a word to read and ask the children to find and write the word you just said.
• Have the children write the words in a column if you want to keep their word as a quiz
• Continue until all the words have been named and written.

individual review or quiz

Use the reverse sides of the SnapWords® cards for assessment.
TEACHER-LED
• Show the child one word at a time.
• The first time through the set, if he hesitates naming a word, turn it over to show him the image side briefly as a reminder.
• On the next time through the words, if a child misses a word, lay it down and keep on going.
• Use the images to review the words he was not able to read instantly.
• When he can read every plain word on sight, he will have gained 10-12 new friends!

PEER-LED
• Pair students so that you are relatively sure one in the pair is fluent with the group of words.
• Have the fluent child be the first to be reviewed in order to give the other child a little review before it is his turn to answer.
• The questioner will hold up a card at a time on the plain side for his partner to read. Follow the same procedure as in the teacher-led activity above.

WORD RECOGNITION - review & assessment

word flip

- Display your SnapWords® groups of cards in the pocket chart with images facing.
- Tell the children that they are going to get to vote for the words they can recognize on sight without the picture showing.
- You will turn over the words the children select so the plain backs are showing.
- To begin, ask for suggestions on which word is the easiest and therefore the word that should be turned over first.
- When you have a consensus, turn that word over with a big flurry.
- Have the children vote on which word to turn over next.
- After you have turned over the third word, ask the children if they can still read the words you turned over.
- When you feel the class can all read all the words in the group from the plain side of the card, celebrate and tell the class you will be bringing out a new group of words!

file folder game

Make games in file folders ahead of time. If you laminate them, you will be able to use them year after year.

MAKE THE GAMES
- Open a file folder flat.
- Plan the winding path the words will take from the starting point to the end.
- Write the sight words you have been learning on the path.
- Each child will take a turn tossing one die. He will read the same number of words as the number of dots on his die.
- Children will take turns until they have read all the words.
- To provide more practice, design the game so the road never ends and have each child start in a different place along the path.
- If desired, add a few twists to the game such as "go back two spaces," or "lose a turn" or "play twice."

READING - oral phrase building

phrase pop up | 11 |

POP UP 2:
• Use a group of SnapWords® cards the children can read.
• This time when you play Pop Up, the children will take turns linking two words into a phrase (for instance, "go up" or "come here.")
• When children are comfortable linking two words, move on to Pop Up 3.

POP UP 3:
• This game requires making a 3 word phrase such as "come help me" or "play with me" or come down here."

Increase the number of words in the phrase as you feel the children are comfortable with the level you just finished.

READING - tactile phrase building

it's a windy day!

Use your group of SnapWords® with the right sides facing the children.

• Create phrases with the SnapWords, with one phrase in each row of the pocket chart.
• Read the phrases together.
• Have the children close their eyes and turn over a card or two.
• Tell the children that the wind blew so hard, it turned over a couple of the words.
• Read the phrases together, challenging the children to read without the pictures showing!
• Have them close their eyes again and repeat.
• When all the words are turned over, just leave them as a display for a couple of days for the children to practice reading.

READING - sentence building

13 | sentence tree

- Use a group of SnapWords® cards the children can read.
- Start in the middle of the left hand side of the pocket chart with the word "I".
- Next, in a column to the right of I, place three verbs such as "want, need, have."
- After each of the verbs, place a plain word "to."
- Finally, use the SnapWords® to finish each sentence.
- Practice reading the sentences, taking turns, turning over words like you did in the first windy day game.

Example of pocket chart with words:

			call my mom.
			play with you.
			run down here.
			work with you.
	need	to	work down here.
I	want	to	sit with you.
	have	to	call you now.
			eat it all.

To read a sentence, begin at the left with the word I, and then pick a verb, and finally a sentence ending. *Don't hesitate to use the terms "pronoun," "verb" and "infinitive" when you are doing sentence building together.*

14 | whole group mixed up sentences

Use your group of SnapWords® with the right side facing the children.

- Create mixed up phrases in the pocket chart. Work together, discussing how to rearrange the words to make the phrase make sense.
- In the beginning, if needed, say the correct phrase and let that verbal guide help the children rearrange the words.
- Later, in a guided activity, let the children collaborate on putting the words in the order that makes sense.
- Much later, use this activity in centers where partners work together to make sentences to unscramble for each other.
- Do this activity with the stylized sides showing at first and then switch to the plain sides as the children are more comfortable with reading the words instantly.

READING and WRITING connection

sentence starters   15

- Put sentence starters in a pocket chart.
- Share with the children that they will be able to choose whichever starter they'd like to finish. They will write the starter on their paper and then decide how they would like to finish the sentence.
- All children should be encouraged to illustrate their work.
- Sentence starter ideas: "I have a little…." "I want to make a….." "Do you want to……….?" "I see a funny, little….."
- Ask children to read their writing to someone so they can practice reading without having the sight word cards.

Increase the number of words in the phrase as you feel the children are comfortable with the level you just finished.

story starters   16

- Create sentences in the pocket chart. Make them as ambiguous as you can. (Ex: "Will you stop that now!" and "I will get the green one.")
- Instruct the children that they may choose one sentence and build a story around the sentence and write what they want it to mean.
- Other ideas could include "There are two up the tree" and "I can see you go in there." The idea is that the child will invent the scenario and write about it and illustrate it. In the first sentence, for example, there are two of what up the tree?
- Always allow for share time, whether whole group or with partners.

READING and WRITING comprehension

 missing words

- Create sentences in the pocket chart, leaving spaces where words are missing.
- Children will select a word to go in the blank space for each sentence, and then will write each sentence on paper.
- All children should be encouraged to illustrate their work.
- Ex: "I want to _____ with you."
- Or "Do you have to _____ now?"

 journal writing

- Display a sentence in the pocket chart.
- Ask the children to draw a picture of what the sentence says.
- Ex: "The funny blue cat sat on me." "Will you come help me work?" "Come down here now!"
- The pictures will show whether or not the child understood the sentence.
- If desired, you could display a handful of sentences and let each child choose the one they want to illustrate.

READING and WRITING comprehension

new word work | 19

- Have a new sight word as the focus for each day.
- Share with the children that it is wonderful they can recognize this word on sight, but now they are going to make sure they know how to use it really well.
- Ask children to write the new word in his or her writing journal, stylize it, and then use it in a sentence.
- Ex: for JUMP, the child will stylize the word as he or she desires, then will write something like this: "My cat can jump like me."

STRUCTURAL ANALYSIS of WORDS

word family introduction | 20

- Select a target sight word for the day.
- Identify a focus spelling pattern (ie: the portion of the word on which you want to focus. (Ex: short "a" sound, or "er" ending, or "oo" as in "soon.")
- Brainstorm other words containing that target spelling.
- Write the words on a whiteboard or chart paper.
- Underline the target spelling in each word or write those letters in a different color.
- Ex: for the sight word "AT," add words such as "cat, fat, sat," etc. For "oo" they might say "moon, noon, boot, zoo," etc.
- Learning is enhanced if children are writing on whiteboards as you generate this list together.

STRUCTURAL ANALYSIS of WORDS

 word families scavenger hunt

- Identify a sound spelling such as "ow" in "cow" from a selected sight word.
- Children will search through books or through the sight word wall until they find more examples of words containing this sound spelling.
- Other "ow" sight words include "down, now, how."
- If the children use books to search for more words, they could work in pairs and write down all the "ow" words they can find.
- Share lists with the class. I posted these lists in my classroom on long strips of paper so the words made a long column. I encouraged the children to add to this list as they encountered other matching words in their reading.

 odd man out

- Choose and display 4 words which have a sound spelling in common and one word that does not match. (Ex: "see, green, three, here.")
- Ask children to study the words.
- The task is for the children to identify which word does NOT belong in the group. In the example given above, "here" does not belong, as it does not have the "ee" spelling.
- In "where, there, here, three," "three" does not belong.
- In "five, like, with," "with" does not belong.

STRUCTURAL ANALYSIS of WORDS

word sort

• Introduce this activity to the whole group, and then you may use it as a center activity.
• Give each group of children about 7-8 cards, and choose words for each set that combine two different target sound spellings. Ex: old, cold, hold, told, now, down, how, brown.
• Combine picture cards with plain cards you have prepared by writing additional words on index cards as needed to make the game work. Shuffle.
• Ask the group to sort the word cards into two piles. They must agree on how to sort their cards and be able to verbalize their choice when they have finished.
• In our example, children would create a stack of "old" cards and one of "ow" cards.

PHONEMIC AWARENESS

word morph

• Choose one sight word for the day. Display in pocket chart.
• Children have markers and whiteboards.
• Identify each sound in the target word. For example, in "not," you would segment /n/, /o/, /t/.
• Have children sound and write "not" on their whiteboards.
• Ask, "Can you change 'not' into 'hot'?"
• Check to see if children are changing the n into an h.
• Continue with letter replacements, such as in "cot, pot, dot."

PHONEMIC AWARENESS

 more word morph

- Continue activity 24, but this time, after the children have written the first word (not) on their whiteboards, ask:
- "Can you make 'not' into 'hot'?"
- "Can you make 'hot' into 'hop'?"
- "Hop" into "mop?"
- "Mop" into "map?"
- "Map" into "tap?"
- "Tap" into "top?"
- "Top" into "stop?"
- One idea is to use the sight word of the day as the game starter each time.

 add a letter

- In this version, you start with a tiny sight word, such as "a."
- Each change to the word requires adding one letter to the previous word.
- "A" could become "at."
- "At" could become "rat."
- "Rat" could become "brat."
- "Brat" could become "brats."

- "I" turns into "it," "it" to "sit," "sit" to "spit," "spit" to "split," "split" to "splits."

List of 607 Child1st SnapWords® from www.child1st.com

List A							Nouns 2	
a	all	around	both	different	hat	few		answer
an	am	before	buy	drink	home	finally	air	became
and	any	began	clean	enough	hope	free	Nouns 1	become
are	ask	better	close	father	later	front	baby	begin
as	ate	bring	could	flapping	letter	fun	ball	being
at	away	came	done	giggle	longer	great	bird	believe
back	be	cold	draw	heard	love	half	boat	broke
big	cut	day	even	hitch	maybe	heavy	body	brought
but	eat	didn't	every	hundred	men	important	boot	build
by	fast	does	fall	husband	money	inside	boy	can't
call	fly	don't	full	imagine	morning	instead	bus	care
can	from	far	goes	indeed	name	large	children	catch
come	funny	find	grow	instant	night	less	city	caught
did	gave	first	high	it's	o'clock	lot, (lots)	clothes	change
do	good	found	hot	mother	order	mad	cloud	complete
down	got	give	hurt	passenger	pair	main	country	couldn't
for	him	giving	I'm	playmate	part	nice	crab	died
get	into	going	keep	pleasant	present	often	Dad	explain
go	its	gone	laugh	please	push	page	desk	feel
has	just	had	leave	pleasure	room	perhaps	dinner	fell
have	last	hard	left	prize	sat	possible	dream	fight
he	let	her	light	realized	second	probably	ears	finish
help	many	hold	mean	shall	seem	quick	earth	fix
here	may	how	might	stove	set	ready	eyes	follow
hi	must	kind	most	struggled	sister	real	fact	forget
his	new	know	myself	stuck	someone	really	family	form
I	of	live	near	stumbled	something	scared	fare	grade
if	our	long	need	thank	special	several	feet	hear
in	pull	made	once	thought	stand	sick	field	hit
is	put	man	only	through	store	side	fire	kept
it	read	more	open	together	such	simple	fish	killed
jump	run	much	pretty	tomorrow	thing	since	flower	knew
like	saw	never	right	toward	third	size	food	learn
little	say	next	round	twice	though	sound	friend	listen
look	she	off	same	wash	until	sure	game	lost
make	show	oh	short	whole	way	that's	girl	mind
me	sing	old	should	willing	yesterday	themselves	grass	miss
my	still	other	sleep	wish	yours	they're	group	move
no	take	over	small	wonderful		top	hair	organize
not	tell	own	start	write	List G		head	quit
now	than	pick	their		able	NCDMS	hill	rest
on	that	ride	today	List F	above	Numbers:	hour	seen
or	them	some	turn	along	against	1-20	house	shot
out	then	soon	upon	also	almost	Colors:	idea	speak
play	they	there	use	anything	already	black	insect	spend
ran	too	these	warm	bed	although	blue	island	stay
said	took	think	well	box	among	brown	kids	stood
see	try	those	while	car	bad	gold	lady	study
sit	us	told	would	cat	beautiful	gray	land	succeed
so	went	under	yet	coat	behind	green	life	talk
stop	what	very		color	below	orange	line	teach
the	when	walk	List E	dear	between	pink	list	throw
this	who	was	accident	dog	brother	purple	lizard	travel
to	why	were	basement	door	certain	red	lunch	tried
up	with	where	because	dress	dark	silver	Mom	understand
want	work	which	bicycle	each	deep	white	moon	wasn't
we	yes		breath	early	dry	yellow	nothing	watch
will	your	List D	careful	end	during	12 Months	number	win
you		across	carry	face	easy	7 Days	ocean	winter	woke
	List C	always	certainly	fat	either	4 Seasons	paper	worry
List B		after	animal	clapped	fine	else	day	park	wouldn't
	again	been	company	hand	ever	worm	party	
about	another	best	decide	happy	favorite	year	past	
							people	
							picture	
							place	
							planet	
							plant	
							problem	
							rain	
							reason	
							river	
							rock	
							sand	
							school	
							sea	
							ship	
							shirt	
							shoe	
							sign	
							sky	
							snake	
							snow	
							space	
							spider	
							spring	
							stake	
							state	
							stick	
							storm	
							story	
							street	
							stuff	
							summer	
							sun	
							table	
							teacher	
							things	
							time	
							town	
							tree	
							trouble	
							water	
							week	
							wind	
							woman	
							words	
							world	
							yard	
							year	
							Verbs	
							add	